MAGNET DOWSING
OR
THE MAGNET STUDY OF LIFE

By

Dr. B. BHATTACHARYYA M.A., Ph. D.,

Rajyaratna, Jnanjyoti.

Reprinted 1985
By

HEALTH RESEARCH

Dr. BENOYTOSH BHATTACHARYYA

(Taken from Biographical Encyclopaedia of the World, 1946)

Born on January 6, 1897, at Deasin, Burdwan Dt.; Son of Mahamahopadhyay Haraprasad Shastri and Hemanta Kumari Devi ; educated at Scottish Churches College, Calcutta, B.A. 1917 ; Calcutta University, M.A. (Gold Medalist) 1919 ; Dacca University first Ph.D. 1925, married Labanya Mayi Devi 1920, two sons and five daughters—Omi, Ranabir, Pratima, Ramola, Shyamola, Chanchala and Vidya. Dr. Bhattacharyya began as Bengal Government Research Scholar, 1920 ; entered Baroda State service as General Editor Gaekwar's Oriental Series, 1924 ; part-time Professor of Sanskrit, Baroda College, 1929 ; in charge of Marathi and Gujarati publications of the State, 1931 ; recognised as Post Graduate Teacher by Bombay University, 1932 ; Examiner for Ph.D. degree 1942, and for Bhagwanlal Indaji Gold Medal and Prize, 1943. Director of Oriental Institute 1927-1952 and published under his general editorship over 80 original works on Indology in Gaekwar's Oriental series. Organised Seventh All India Oriental Conference 1933 (Local Secretary); Local Secretary 17th. Bengal Literary Conference, 1922. Has established charitable dispensaries for Homoeopathy since 1927. Life Member of Royal Asiatic Society of Bengal, Calcutta, and of Bhandarkar Oriental Research Institute, Poona. Member or sometime Member of Bihar and Orissa Research Institute, Patna, Vangiya Sahitya Parishad, Calcutta ; International Psychic Research Congress, London ; and Kern Institute, Leiden, Holland. Received title of Rajyaratna and Gold Medal ; title of Jnanjyoti and a Silver Medal ; Diamond Jubilee Medal ; Clubs : Sayaji Vihar, Baroda ; Bengali, Delhi ; University Institute, Calcutta. Hobbies : Photography, astrology, mystic sciences, Homoeopathy, sculpture and painting. Publications : Indian Buddhist Iconography, 1924 ; An Introduction to Buddhist Esoterism ; 1932 ; Proceedings and Transactions of the Seventh Indian Oriental Conference 1935 ; Sadhanmala, 1925, 1928 ; also over 200 original papers and articles.

His natural aptitude to explore the hidden powers of Nature for the service of Humanity developed in him an insight into the therapeutic value of Gems and Magnets ; did extensive research in those fields ; has published the result of his experience in his books Science of Cosmic Ray Therapy, Gem Therapy, Magnet Dowsing, Science of Tridosha, Septenate Mixtures in Homoeopathy etc. Has trained students on this novel method of curing disease by Teletherapy. A number of Teletherapy centres were opened under his guidance.

After retiring from Baroda Service in 1952 he was continuing his research work at Shastri Villa, Naihati. The end of his busy life came on June 22, 1964.

PREFACE

Shri K. L. Mukhopadhyay informed me sometime in 1965 that the stock of *Magnet Dowsing* was low and requested me go ahead with the revision of the book for a new edition. My revered father left this world unexpectedly on 22nd June, 1964 after a brief illness. His sudden departure put a great burden on my feeble shoulders and I found myself in a quandary. My father's Teletherapy cases were to be handled as also new cases were to be examined. Whatever time was left was consumed by a heavy correspondence inland and overseas. For all these reasons I could not undertake the work of revising this book. But when Shri K. L. Mukhopadhyay informed me that the book was out of print, I left all work and started on this.

It is a very difficult task to revise a book written by a great scholar like my father but I had to undertake it, as there was nobody else who would do it. There may be many shortcomings but I hope the readers of this book will forgive this. My purpose in revising this book was to give information about new developments that have taken place since the last publication. As far as possible, I have maintained the same text except for some changes here and there, and some new information which were not there in the first edition.

My grateful acknowledgments are due to several friends who have taught me many new things. My special thanks are due to my esteemed friend Dr. Albert Roy Davis, who taught me the mysteries of the bar magnet. My grateful thanks are also due to my friend and colleague Dr. H. S. Harnal who has supplied me with many charts all of which I could not include in the book for want of space. I have included several of them which I found to be of special interest.

I also owe my thanks and indebtedness to my dear friend Shri Krishna Chandra Chakravarty who has helped me in every way for the completion of this book.

I am also grateful to my uncle Prof. Manjugopal Bhattacharyya for going through the manuscript but for which many more mistakes would have crept in.

I cannot forget to mention the name of Shri Ashim Kumar Chakravarty, a bright young man with a future, who has beautifully prepared some of the new charts included in this book.

Lastly, but for Shri K. L. Mukhopadhyay this book would never have seen daylight. I am ever grateful and indebted to him for extending to me the same help that he gave my father.

To all those who helped me but whose names I have not mentioned here I tender my most grateful acknowledgments.

A. K. Bhattacharya

Shastri Villa,
Naihati,
Sreepanchami, February 14, 1967.

LIST OF ILLUSTRATIONS

CONTENTS

INTRODUCTION

2nd. Edition of Magnet Dowsing

The first edition of this book came out in 1960 and the end of 1966 saw the 'Out of Print' notice at the publisher's end. This in itself is very remarkable, as it showed that many people, who did not know what Radiesthesia was, were actually working on its principles. The remarkable thing about the present book is that it is, to my knowledge, the only book on the subject of Radiesthesia written and published in India. There are many good books published in England, France, America and Germany, but as they are not easily available here, this book has served the purpose of satisfying the curiosity of the inquirer.

Even though the main principles have not changed much since the publication of the first edition, still some important additions have been made in the science during the last seven years. Also, I feel that some of the opinions of the Western authors on this subject should be known by the readers of this book. Some changes have been made in the original text of the book, and the last chapter on Advanced Dowsing has practically been rewritten in view of the advances made in the last seven years.

Firstly, I find that some people get confused with the two words 'Radiesthesia' and 'Dowsing'. Radiesthesia is the science dealing with the study of energy fields by sensitive individuals or the study of the human body's perception of, and reaction to ultra-low intensity radiation. The operational technique is known by various terms, such as, dowsing, divining, etc., which means the same thing. Radiesthesia mostly deals with detecting diseases in human beings or is applied in medical divining, whereas Dowsing deals with underground water, oil, minerals, etc. The operation when applied to health, is termed as Radiesthesia in England and Radionics in America.

Secondly, there seems to be much confusion as regards the actual pendulum movement and its cause. I am giving below the opinions of some of the English and American authors about the actual cause or reason for the movement of the pendulum.

Dr. Henry Tomlinson, in his book "The Divination of Disease" explains as follows. There appear to be two sources of energy which produce the effects registered by the pendulum, or two schools of thought regarding it. One is that the energy is a part of the electro-magnetic band, and the human body is a mass of various electrical forces. The other is that dowsing effects are metaphysical in nature, a product of the mind.

Mr. W. H. Trinder in his book "Dowsing" writes, "Dowsers hold that all matter gives off a ray which causes a change of tone in certain muscles of the fore-arm. This causes a reflex action of the muscles which move the dowsing instrument without the individual's volition, the reflex action not being felt by him."

Mr. D. W. Wethered writes about dowsing in his book "Health and Pendulum", "The word *radiation* generally implies to the layman intangible, non-material emanations given off by certain bodies. But to the physicist it means energy existing in the electro-magnetic waves. The word *influences* is used to denote non-material forces as detected by the pendulum, and radiation indicates where the forces operate. Human sensitivity to these forces is registered by the pendulum. The human being is surrounded by rings or shells of influence, where the pendulum reacts, and which may be assumed to be comparable to magnetic fields existing in electrical phenomena. It is a complex broadcasting station giving off different radiations from every tissue, organ and structure. The pendulum deals with the radiations thus detected and magnifies the effect of the radiations on the nervo-mascular system of the operator."

Mrs. Marguerite Maury in her book "How to Dowse" is of the opinion that the sensitivity to external influences revealed by movements of a pendulum is physical or material. She continues that although the biological mechanism by which pendulum movements are controlled is unknown, they can be reduced to a system and put to real use.

Mr. Max Freedom Long, in his book "The Secret Science at Work" says that dowsing is a technique for using the pendulum as a means of communication with the subconscious.

Dr. R. R. Weiss asserts in his book "How to Use a Pendulum" that subconscious mind actually controls the swinging of the pendulum. When the subconscious mind has a message to convey it moves the arm of the pendulum operator causing the pendulum to swing and indicate a YES or NO, or indifferent answer, *i.e.* positive, negative or neutral.

I think that the movement of the pendulum is due to the following causes. When the pendulum is held suspended by a thread with the thumb and the forefinger, the di-electric energy of the operator passes through the string and collects in the pendulum and thereby a magnetic field is created which is dynamic in character. This magnetic field becomes the omniscient field. When a question is put to the Overself which has all the answers to past, present and future questions, there are subtle changes in the electrical frequencies causing the pendulum to move in a certain fashion. The different types of movements are described in the text in the chapter on "Dowsing Technique."

Opinions may differ from person to person on the actual cause of movement of the pendule, but one thing is certain that it does move. The first known case of such movement was in the hands of Moses. When he was moving in the desert with a hazel twig with the thought of water in his mind, the twig suddenly hit a rock under which water was found. Instances may also be cited from modern times which will prove that when modern science which is so very advanced today failed, the dowser succeeded. The Secretary of Mahatma Gandhi, Mr. Sudhir Ghose has narrated a very interesting story in the Bengali Weekly "Desh" of 14. 1. 67. He writes that during the planning of the township of Faridabad the engineers informed him that there was no water underground. When it was brought to the notice of Pandit Jawaharlal Nehru, the then Prime Minister of India, he advised Mr. Ghose to contact a water-diviner who was living in the State of Nawanagar in Gujarat. A tall man known as 'Paniwalla Maharaj', arrived and moving over the 3000 acres of land pointed out ten places, where he claimed there was water underground. The engineers who were present laughed it out as a highly unscientific and fantastic procedure and waited to see the result. To the amazement of all, it was found that all the ten places which were drilled were overflowing with water.

The above incident shows that, even though we may not be able to give a proper scientific explanation for the art of dowsing the fact remains that it works even when modern scientific methods totally fail.

The next point of importance is the choice of pendulums. Here also, there are differing opinions. Some prefer only neutral articles to be used as pendulums, such as, wood, glass, ivory, crystal and silver. Others have used a bead, a button, a ring or any other article for dowsing purpose with good results. I have used magnets, crystals, silver, gems, etc. with equally good results. It is not the pendule but the person behind the pendule that is important. Another point is the length of suspension of the pendule. My revered father used to suspend the pendule to a length of about 6" which I also follow in my dowsing work. Others claim that a height of about 2½" to 3" is ideal. This is a matter of personal convenience and hence in my opinion it is not of vital importance. The pendule should be suspended to a length that gives easy movement.

Preliminaries being over, I shall now discuss the actual art of Radiesthesia. Principally, for dowsing purpose, the requirements are only a pendule, a table and a chair to sit on. This is explained in detail inside in the text and hence I will not elaborate on the rules of dowsing. My father has stated before, in the introduction to the first edition, that because he found the pendule works on the principle of magnetism, he wanted to make the field more powerful with the help of permanent magnets. He has already described the first magnet box as was prepared by him in the first edition and hence, I shall deal with the later developments.

The need for improvement in the technique arose because

in spite of this magnet box the results obtained by others practising this science did not coincide. A new instrument was constructed in which 24 power 'C' Eclipse magnets were used. The new instrument was divided into three sections. In the left hand section 12 magnets were fixed to iron plates with their North Pole facing the top, and in the right hand section the other 12 magnets were arranged with their South Pole facing the top. The central portion was the neutral area where the dowsing was actually done. For a time it was found that no further change would be necessary as this instrument would answer the need for accurate dowsing results. But, again it was felt that something was wrong somewhere, as the results obtained by several persons were found to be unsatisfactory. The reason was that the neutral area though highly magnetised, was static in action and to make it dynamic the di-electric energy of the operator was of prime importance. Operators with a generous supply of di-electric energy could get very good results but the results obtained by persons whose di-electric energy was low were faulty and defective. During this time a very good and enthusiastic friend of my father Mr. T. H. Vaswani, who was a wireless engineer, suggested the use of electro-magnets instead of permanent magnets. He prepared an instrument with 3 electro-magnets which were connected in a series connection giving approximately 80 volts each. Immediately it was found that this was the thing that was needed for eliminating the individual's di-electric energy problem. As the electro-magnets had the function of changing polarity 50 times as per the electrical line cycle, the field became dynamic in action. Hence, when the operator started dowsing on this field, there was no additional need for him to make the field dynamic with his own di-electric energy. Many types of electromagnets were experimented upon to get stronger magnetic fields, and I am glad to say that at last we have been able to get electro-magnets to our entire satisfaction. This instrument gives more accurate results even with average dowsers, who were unable to get them with the old instrument. I have described the instrument in detail in the text and hence will not repeat it here.

After going through many books on Radiesthesia and practising the art myself for the last 17 years, I have come to the conclusion that this is an occult science and all the rules and regulations regarding the practice of an occult science have to be observed, if one expects to gain efficiency in the art. The first essential rule is that this science should never be practised for personal gain or for injuring others. The answers gained from dowsing should never be communicated to others, as they are secrets revealed to the operator by his overself and hence should be kept a secret. One may for practice ask ordinary questions *e.g.* as for instance whether a person would come on a certain day, or would there be rain on a particular day, and questions of this sort, but once the operator is convinced about his ability to dowse, he should refrain from such frivolous dowsing. As far as possible, demonstrations should not be given in public, as this is not a plaything but a very serious investigating instrument. Most of the Western dowsers agree on this point. It must definitely be borne in mind that the science of Radiesthesia is not for those small minded men who are led by a selfish greed for money to practise it. For them it will be merely a waste of time.

All the charts given in the 1st edition are included in this book and some more charts have been added for the benefit of all. Many more charts can be added but I shall repeat the words of my father that there will be no end to them. Charts may be drawn according to the requirements of individuals as shown in this book. When there is no chart, psychic dowsing may be practised on the magnet box.

The science of magnet dowsing is still in its infancy. There is scope for improvement regarding its technique ; the implements may also have to be changed or modified. A vast field for research is being thrown open to all who wish to undertake it. This is the science of sciences transcending every other science in this world. It is hoped that people interested in searching for Truth will undertake this in large numbers. This being an occult science it will be of great help to those who wish to apply its benefits for the good of others. What more good one can do than give relief to a suffering mortal ?

INTRODUCTION

A glimpse of the future, a knowledge of the unknown, however, dim and indefinite it may be, is prized by a section of thinkers in all parts of the world. A tremendous craving of the human mind for the supersensory cosmic vibrations gives rise to various sciences and pseudo-sciences including divination or dowsing. Radi-esthesia as practised in modern times is one form of dowsing. Knowledge derived from dowsing means power over the elements and with this knowledge many good and bad things can be done, and had been done in the past.

But when we look around there is not a good and reliable book on divination or dowsing. On the subject of Radionics or Radi-esthesia indeed there are many good books published in European countries and America. Now-a-days it is believed that Homoeopathy cannot be efficiently practised without a good knowledge of Radi-esthesia, and Homoeopaths of every country except India are eagerly learning and practising this wonderful method. To have a good library of books on Radi-esthesia will mean a considerable sum running to several thousands, but alas ! in India the word Radi-esthesia is not even heard and it is as unfamiliar to-day as the Chinese Ouizi.

All along I had an intense desire to learn the occult sciences, such as Palmistry, Astrology, Graphology, Colour Therapy, Ayurveda, Tridosha and the rest, at various periods of my life. In the year 1948 I came in contact with a brilliant American scholar, Mr. Howard D. Stangle of New York, who appeared to me to be a versatile genius. He taught me the rudiments of Radi-esthesia which I started practising almost immediately with the simplest possible appliances such as a small table and a small crystal pendule weighing half an ounce and a black string about 3″ inches long with which the pendule used to be held in the right hand. With these simple devices I made thousands of tests and I was satisfied with the result. It was a great joy to me to know, albeit partially, all the problems which defied an answer before I learnt Radi-esthesia.

Gradually, it dawned on me that the pendule works on the principle of magnetism and that the magnetic currents rising from the feet to the whole body concentrate on the little pendule in the right hand, and because there is no other channel for escape the currents condense themselves round the pendule and thus form a magnetic field. Omniscience resides in this magnetic field, and this omniscience in the magnetic field is exploited by Radi-esthesia. Thus the problem arose in my mind as to how this magnetic field can be expanded and made more powerful so as to make it more reliable and more accurate.

Later on, Mr. Stangle drew my attention to the wonderful healing properties of the magnet, and this opened up a new series of experiments for discovering the healing properties of the magnet following the teachings of Paracelsus. In these experiments I was fairly successful, but I discovered that my interests were not there in spite of the success achieved by me in a brief space of time.

Still later, I argued in my mind that if Radi-esthesia tests are dependent on magnetism, why not use the magnets themselves for generating magnetic power. Thus I was led to the threshold of an entirely new system of magnet dowsing following the principles of Radi-esthesia and altering them a bit to suit the convenience of this new method.

Instead of the usual pendule of Radi-esthesia I started using the horse-shoe magnet in the right hand held by the thumb and the first finger and suspended by a black cord 3" inches long. As I went on with experiments on various objects and photographs, I further discovered that the results were not constant and that they varied even at one sitting. At different sittings the result of tests almost always varied.

In order to find out the reason for this variation in result a search had to be made into our scriptures which definitely laid down that a *Vidya* (occult practice) should not be practised on certain days of the lunar month. For instance, the New Moon days, the Full Moon days, the first, eighth and the fourteenth days of the lunar months should be avoided for such practices, indicating thereby that results obtained on these days are likely to be faulty or erroneous. It appeared very reasonable because the magnetic currents released by the Sun and the Moon have powerful effects on this Planet, and as such when the currents released by these two great luminaries are at variance and inharmonious, tests with magnets are likely to be vitiated. When, for instance, the Sun and the Moon are in opposition, square or conjunction inharmonious vibrations are seen at play, and this is the chief reason why magnetic tests should not be attempted on such days and circumstances. Thus the magnet practice had to be confined to the few days that were available after the elimination of objectionable days.

In spite of this the tests varied although not as frequently as before, but the results were not reliable. This led to further magnetisation of the body and mind by holding a magnet in the left hand at the time of conducting magnetic tests. This new magnetisation had marked effect on tests and the pendule did not move as I wished in my conscious or subconscious mind. It gave better results and especially in Teletherapy cosmic colours for individual patients could be selected with ease and accuracy. Still there were cases where the selected colour was not effective. Thinking that the method of magnetisation is still defective further intensification of magnetic power was done by designing what is known as the Magnet Box, a description of which will appear in the sequel in the last chapter on Advanced Dowsing. The box contains sixteen powerful horse-shoe magnets arranged in a particular manner so as to deflect the compass, when placed on the magnetic flap, to the extent of 90 to 100 degrees to the right.

The flap covering the magnets in the magnet box serves as a powerful magnetic field which magnetises anything and everything placed on it. The magnet box thus completes the limit in magnetisation, and the results of tests obtained on this magnet box were found to be more accurate than those obtained hitherto. The colours selected on this magnet box for Teletherapy become at once effective, and cure was effected in a series of bad patients by pinning their signatures on to the walls of the radiating cabinets.

On this flap of the magnet box any photograph examined with the help of the Percentage Chart showed the percentage of health of individual patients rather accurately, and whenever the percentage went beyond 95% or reached 100% the patient in question was to die soon or was awaiting death. Nothing can be done to help him even if all the systems of treatment were tried one after another. On the other hand even when the case was serious or apparently hopeless, the patient recovered completely when the percentage showed anywhere near or below 75%. Cosmic colour radiation by teletherapy shows excellent results when confirmed by magnet dowsing.

A very generous friend in England presented to me a beautiful book on Radi-esthesia which gave a full account of what is known as the Rule Test in Radi-esthesia. This Rule Test was entirely new but its results were attractive, and had to be practised for a long time before it could be adjusted to the system of magnet dowsing. Thus two magnet boxes as described before, were prepared with sixteen magnets in each, in order that the two ends of the 100 CM Rule magnetised by resting the two terminals on the two magnet boxes. Containers were then attached to the two ends for keeping different materials for the purpose of comparison and tests.

The Rule Test is found to be essential where high grade of accuracy is a desideratum. Thus when before radiating a particular cosmic ray to a photograph it is found desirable to test the photograph of the patient against all the nine gems before ascertaining the true colour hunger, because in teletherapy correct colour hunger is the true cause of the disease. On one container should be kept the photograph of the patient. On another container one after another all the nine gems should be placed for the purpose of test. For instance, a Ruby should first be placed in the other container, and in order to find out its suitability for the patient in the first container, a small pendule should be operated on the middle point of the Rule at approximately 50 CM. If the pendule moves up and down then the gem colour of the Ruby which is Red is the right colour for the purpose of radiation. But on the other hand if the pendule moves parallel to the Rule then the gem colour in question is incorrect and should be rejected. It is in this way after careful scrutiny and testing with the Rule and the gems the colour should be fixed, and then radiated to the patient wherever he may be in any part of the world for curative result.

Here a few words are necessary to explain the origin of the cosmic ray therapy or Teletherapy.[1] Mr. Howard D. Stangle who has been referred to earlier suggested to me that the simplest way of broadcasting Homoeopathic remedies to a patient far away or near is to hold the medicine phial in the left hand and over the photograph of the patient rotate the pendule as long as desired or as long as possible. Many experiments were made on this line, and although the method was right, very little could be achieved except to silence a crying child or stop a diarrhoea by this method. Medicine broadcast was tried in the case of a dear cancer patient but all the efforts proved of no avail, although a doctor friend in Australia was definitely in favour of this method of broadcast in all incurable cases.

Later, when our favourite kitten was seriously ill and was on the point of death this direct method was tried. The crystal pendule was taken in the right hand; it was rotated from the left to the right over the head of the dying kitten which was lying motionless. When nearly one thousand rotations were complete the paralysed hair of the head began

---

1. All informations about the science and practice of Teletherapy will be found in the work TELETHERAPY, a collection of articles on Teletherapy. The 2nd. edition of Science of Cosmic Ray Therapy is under compilation.

rising one after another very slowly. This was evidently an encouraging sign and the rotations were continued until the whole head of the kitten became normal and fully puffed up. When 2000 rotations of the pendule were completed the whole kitten from head to tail became puffed and appeared quite lively. Then the kitten tried to move and fell down several times until after a few minutes was strong enough to walk to the edge of the cot and from there it jumped to the ground. The kitten got cured of its ailment and lived thereafter for five long years.

The rotation of the pendule whether slow or fast, releases certain life-giving forces from the atmosphere which can be employed for human good, and of this I became pretty sure after the kitten incident. But the rotation of the pendule by hand for thousands of times is a difficult and tiresome affair, and that is why some kind of mechanised method became necessary. Mr. Stangle suggested the fan motor. Thereafter, the motor of an old fashioned discarded table fan was taken, and a disc was fitted to its shaft containing four crystal balls. When the motor started rotating it was a mechanised Radiesthesia. With electricity the motor and the disc moved furiously and imparted a flood of radiation to the objects kept before or under the disc, be it kitten, photograph, oil or sugar of milk. All receive the gem vibrations and absorb them equally. That was the beginning of Teletheraphy or the Cosmic Ray Therapy.

The next stage was to employ other gems when crystals failed to give relief, and later, all the seven gems with the exception of Infra-Red and Ultra-Violet. Later on, the diseases had to be classified according to the Five Elements and the Five Vayus, and still later the gems were connected with the Five Elements and the Five *Vayus*. Thus Ruby and Coral came under the element of Fire and the Samana Vayu, Pearl and Diamond were placed under the element of Water and the Apana Vayu, Emerald was placed under the element of Earth and the Prana Vayu, Saphire came under the element of Air and the Udana Vayu, while the Moonstone or Pukhraj was included under the element of Ether and the Vyana Vayu. Once the true character of the gems are known treatment by Teletherapy becomes easy. But to know the true character of gems magnet dowsing is essential.

How important this knowledge derived from magnet dowsing is, may be illustrated by a few examples. In a T.B. case, for instance, it is not difficult to surmise that the disease belongs to the element of Earth and the Prana Vayu. Magnet dowsing will confirm this when the photograph of a patient is tested. Such a patient will show considerable improvement when he receives radiation from a set of Emeralds by Teletherapy. A case of Bronchial Asthma likewise belongs to the element of Water and the Apana Vayu. Thus he requires the vibrations of Orange which is of the nature of Water and Apana Vayu. Radiation of Orange with a set of Pearls will show considerable improvement, although there will be cases requiring Red vibrations of Ruby. These distinctions can be known by consulting the magnet. Weak brained boys and girls and morons usually require Red vibrations because they belong to the element of Fire and the Samana Vayu. People suffering from Sciatica and Neuritis belong to the element of Air and the Udana Vayu. Such people are considerably helped by Violet radiation with a set of Sapphires. Lastly, patients suffering from tumours, swellings and water accumulation, and heart diseases of various kinds, belong to the element of Ether and the Vyana Vayu. Such patients are considerably benefited by the vibrations produced by a set of Moonstones, Topaz or Crystals. When death is approaching and no treatment is effective Moonstone vibrations should be radiated. It is within its power to grant or prolong life, and this is mentioned here lest it should be forgotten in an emergency.

Teletherapy is not possible without a knowledge of Magnet Dowsing. At every step an operator is confounded without the guidance of the magnet. The magnet should be consulted freely in the manner described in the book and all doubts should be first cleared before taking up actual treatment. It will be infinitely better if homoeopathic remedies suitable for a patient are first kept on the magnet box and tested by magnets to find out whether the medicine has the greatest affinity with the patient's disease and symptoms. Even with the help of the Percentage Chart it will be possible to find out whether and to what extent each medicine is agreeable to the patient in question, and of course, unless the medicine is 100% correct, it should not be administered. The simplest method of testing a medicine is to think about the patient and then run the pendule in the right hand against phials of medicine. If the pendule is attracted towards the phial then it can be administered, but if the pendule runs parallel to the phial then it is not the right medicine and will possibly do harm. By this simple method many unnecessary prescriptions can be avoided. A bead, or a *Rudraksa* or a small silver ball may be used to serve as a pendule for this purpose, as indeed for other purposes.

A large number of charts have been used in this brochure, but here it may be remarked that the number of charts can be endless to serve endless purposes of mankind. Man's curiosity knows no bounds and he wants an answer to all his queer problems. Besides the charts hereinafter mentioned and illustrated, hundreds of other charts can be devised to serve as a standard of comparison for magnet dowsing. A gem, for instance, is brought to the dowser for examination as to whether it is genuine or fake. A chart for this is not to be found in this book, but it can be prepared instantly. A piece of white paper or a card may be taken. On it a semi-circle may be drawn which may be divided into two equal parts. The right part should have the inscription "Genuine" while the left should have the word "Fake" On the centre of the base line of the chart the gem should be placed. The pendule should be held over the gem and operated on the magnet box. Then a gentle rotatory movement should be given to the pendule and then allowed to move by its own force. If the pendule goes to the right the gem is genuine, but if it goes left it is fake. But such experiments should be repeated several times before arriving at a decision, and such experiments should always be done on a strong and powerful magnet box. Otherwise it is quite possible some of the tests will go wrong.

Of all studies the study of Man is the greatest and of the highest importance. Man is a small universe, he is a microcosm, and all the forces that are operating in the universe also operate in every man. Man is thus Divinity Itself and therefore Man has to be studied with care and very thoroughly in relation to all the cosmic forces known to mankind, and in this study Magnet Dowsing should play an important part.

Lastly, it should be pointed out here that in this book frequent references have been made to the north and south poles of the magnet. In fixing the poles of a magnet north has been taken in this book as the geographical north, while the word south denotes the geographical south. According to scientists there is a difference between the geographical north and the magnetic north, and the equation is somewhat like this :

Geographical North = Magnetic South
Geographical South = Magnetic North

As it is difficult for the present author to think in terms of the opposites, only geographical directions of the compass have been followed in all instances.

The compass shows one of the needle points always directed towards north or what is called the geographical north. This is what is called the magnetic south by the scientists. But when the compass is held vertically the north point of the needle is attracted towards the meridian, while the south point of the needle is attracted towards the nadir. Does it mean then that the meridian should also be called magnetic south, and the nadir as magnetic north ? This problem has not been properly discussed in the text books on Magnetism I had the good fortune in reading. The compass shows the geographical north and the geographical south. The pole of a magnet which attracts the north point is taken here as the north pole of the magnet and *vice versa*. This nomenclature has been adopted here, and it appears throughout the book, and is required to be noted at the very outset.

CHAPTER I

The Magnet

Even a child knows what a magnet is and how it attracts iron filings, pins, nails and weights made of iron. Scientists have offered many explanations for this remarkable phenomenon displayed by the magnet, and they have found out ways to measure mathematically the power and strength of the individual magnets. Iron can be magnetised by passing electricity through it for some time by a special process. Thus magnetism and electricity are allied sciences.

According to some authorities magnet is so-called because the shepherd Magnes first discovered 2500 years ago that a rock of naturally occuring magnet lodestone, had a strong attraction for his iron crook. Some others have said that the word 'magnet' is derived from Magnesia in which lodestone was first discovered. The scientific name of the magnet is lodestone or loadstone which occurs naturally and has the power to attract iron.

The most conspicuous property of lodestone, namely, its attraction for iron appears to have been familiar to the Greeks as early as 800 B.C. and is mentioned by Homer, Plato, Aristotle and others. Lucretius (96-55 B.C.) stated that the stone can support a chain of little rings each adhering to the one above, and it thus indicates that in his time the phenomenon of magnetisation by induction was also known.

The property of orientation in virtue of which a freely suspended magnet points approximately to the geographical north and south is not referred to by any European writer before the 12th century, though it is said to have been known to the Chinese at a much earlier period. The application of this property to the construction of the mariner's compass is obvious, and it is in connection with navigation that the first reference to it occurs.

Magnets are sold in the market in various shapes and sizes and in various strength, sometimes in the form of a bar, a circular disc, and in horse-shoe shape. This last is useful in dowsing. Magnets are also being used by magicians of repute for many of their uncanny shows, and horse-shoe magnets are widely used in display for the purpose of advertisement.

There are many stories[1] current about the powers of the magnet to which the power to heal the sick is attributed. The famous mystic of Europe, Paracelsus, has given some of the characteristic powers of the magnet, particularly for healing the sick, and there are practical thinkers in America who are carrying on vigorous and systematic researches to discover the various healing properties of the magnet.

A surprising variety of curative powers and other properties are attributed to the lodestone. It is said that lodestone with honey is a purgative medicine and helps elimination of faecal matters. It is also believed that a magnet can draw pain out of the body when properly applied. The magnet is also recommended as an amulet for headaches. Indian gem experts consider the magnet as one of the many

1. The book entitled, *Magnetism* by David Shoenberg in Sigma, Introduction to Science, No. 12 proved of great value to me for understanding the properties of the magnet, and its uses in history, its origin, and the various sciences and superstitions connected with the magnet. I am thankful to the author.

gems which should be worn for various purposes, for health, for prosperity, and for peaceful living generally.

It was also believed that wounds made with a weapon which has been magnetised with a lodestone caused no pain. Lodestone was also recommended for the cure of gout, dropsy, and hernia. Some people believed that lodestone was a poison, but against this there is the instance of the King of Zeilan who had all his meat served in dishes of lodestone, and conceived thereby that he preserved the vigour of his youth. It was also believed that a wife's infidelity might be discovered by putting a magnet under the pillow which prevented her from staying in bed with her husband.

Besides the valuable experiments of Paracelsus, the Homoeopaths of repute have not been slow in discovering and recognising the wonderful healing powers of the magnet. In Homoeopathy the magnet, its south pole and the north pole have been potentized, and these three magnet medicines have exhibited wonderful power over certain series of diseases.

The celebrated Homoeopathic author Dr. Allan in his indispensable book : *The Materia Medica of the Nosodes* has referred extensively to the healing properties of the three medicines prepared from the magnet. The first is Magnetis Poli Ambo, the second is the Magnetis Polus Arcticus, while the third is the Magnetis Polus Australis. The first is the magnet with the two poles potentized, the second is the north pole potentized, while the third is the south pole potentized. All these three magnet medicines are available in various potencies, and they are a great asset to the Homoeopathic healing art, although it must be said that enough attention has not been given to these medicines as they deserved.[1]

These Homoeopathic magnet medicines can create, and cure a host of symptoms and diseases, and they have their definite aggravations and ameliorations like any other Homoeopathic medicine. According to Dr. Allan all the three magnet medicines can cure a host of diseases of the various organs of the human body. Those who are interested in these magnet diseases and magnet medicines are referred to his excellent book already cited (p. 206).[2] This is not the place to discuss the matter more fully.

Magnets have been used by the present writer for reducing swellings and pains, and in such cases the magnet has demonstrated its wonderful power. Mr. Howard D. Stangle in his chapter on the Healing Power of the Magnet has referred to several of the curative properties of the magnet, and has quoted some instances in which the magnet demonstrated its curative powers. The so-called rational scientists have declared without reserve that the magnet has no healing property, and the belief in its powers is nothing but superstition. The healing power of the magnet is proved both by observation and experiment, and as such the opinion of the so-called rational scientists cannot but be accepted with great caution.

The magnet has two poles, north and south. The north pole repels the north of another magnet, but attracts with

1. Dr. H. C. Allen, M. D. in his historical work: *The Materia Medica of the Nosodes with provings of X-Ray*, published by Messrs. Boericke and Tafel, 1910 has not only given the provings, symptoms, aggravations and ameliorations of electricity and X-Ray, but also made elaborate observations on the three magnet medicines referred to above.

2. The descriptions of symptoms cover 50 pages in print. Readers will find here much that is new. Magnet students will do well to study the most valuable observations of Dr. H. C. Allen.

force its southern pole. Similarly, the south pole of a magnet repels the south of another magnet, but attracts forcibly its north pole. The selective power of attraction and repulsion exhibited by the magnet seems to point out the presence of some intelligence in the magnet, however feeble it may be. This intelligence can be and should be exploited for human good.

Also, the two poles of the magnet act differently on human beings, animals and plants. The south pole being hot in character is energising in nature, while the north pole, which is cold in character has a retarding effect. Experiments carried out by Dr. Roy Davis (Green Cove Springs, Florida, U.S.A.) on white mice are worth mentioning here. White mice were implanted with cancer cells and were allowed to grow. When the manifestations were seen experiments were conducted by the application of the two poles, separately. With the application of the north pole of the magnet it was found that the cancerous tumours gradually shrunk in size and ultimately disappeared totally. But, the exact opposite effect was noticed when the south pole of the magnet was applied. The tumours increased rapidly killing the mice. This opens up a vast field of research with the magnets, both permanent and electromagnets, on various diseases to which men are subject.

When a magnet is placed on a table its magnetic field spreads far and wide, and the field can be detected by an ordinary compass. The compass may be moved slowly from short to long distances in front of the two poles of a horse shoe magnet, and when this is done the sensitive needle of the compass will fluctuate and behave differently being influenced by the magnetic power spread out from the magnet. If the magnet is removed from the original place the compass will begin behaving normally by gradually turning to its normal direction, north. More sensitive compasses will show the influence of the magnet from a greater distance and in a feeble manner the magnet throws off its magnetism to long distances. This magnetic field also has a great deal of influence on human beings in correcting their disorders as well as maintaining a sound health. Dr. K. E. Maclean of New York City has been using strong magnetic fields in the treatment of advanced cancer cases, as reported by Joseph F. Goodavage in the magazine 'Fate' in its July, 1964 issue. Results are said to be remarkable. Dr. Maclean at 64 is said to look as if he is 45 after exposing himself to 3600 gauss magnetic field for about 5 years. Dr. Maclean's application of a magnetic field seems to "relieve pain from any cause and to help the human organism to return to normalcy. It is effective in treating most types of Arthritis, Rheumatism and Bursitis."

This magnetic field can be exploited for dowsing purposes, and it is possible to get an answer to all kinds of questions asked to it. With the help of the magnetic field super-sensory cosmic vibrations of all kinds can be detected. How the magnetic field can be exploited for human good will be shown in the sequel. But it must be pointed out in the very beginning that the answers obtained from the magnetic field are to be taken as probabilities only and not certainties, although in many cases whatever is pointed out by the magnet will prove to be absolutely true. As far as possible the operator should know super-sensory vibrations for his own information and not for advertising purposes. The information obtained by magnet dowsing should not be communicated to anybody under ordinary circumstances, otherwise the operator is likely to lose efficiency. Most Western dowsers are agreed on this point.

Again, the magnet should never be used for any base or sinister purpose, or with the sole aim of making money. Magnet Dowsing is a noble science and it should be practised only by those disinterested persons who are urged by a philanthropic motive of giving relief to the ailing and the distressed. For others it will be merely labour wasted as they will never meet with success.

It has been pointed out earlier that the most important study of mankind is man. Man has to be studied from various angles before he can be properly understood. In the matter of disease and medicine, and in the matter of life and death he is also to be studied from different angles, because without these studies neither the disease nor the medicine, neither life nor death can be understood for practical purposes. It should be known, firstly whether a person is attacked with a disease of the earth element or the water element, whether it has ruined the Prana Vayu or the Samana Vayu, and whether the seat of the disease is in the Ajna Chakra or the Manipura Chakra. These and other myriads of questions constitute the problems of life and will be answered by the magnet system advocated in this book. It is the earnest hope of the present writer that the study of man will be further implemented by magnet dowsing, and our knowledge of man and life in general will be more complete than has been possible hitherto. The system must be tried before it is understood, and continuous Sadhana and practice will make the neophyte successful and efficient.

Magnets for healing purpose are available in India from Permanent Magnets Ltd., 20, Old Custom House Road, Bombay-1.

CHAPTER II

A Small Universe

It has been pointed out already that the magnet manifests certain powers of attraction and repulsion. Attraction is equivalent to love, while repulsion is hatred. These two sentiments are clearly demonstrated by the two poles of the magnet. Wherefrom the magnet derives these seemingly wonderful powers is not known to science, while its powers are also only imperfectly known. It is thus necessary to discuss what a magnet is about.

The magnet has two poles, north and south. North attracts the south of another magnet, while it repels the north of another magnet. The south likewise attracts the north and repels the south of another magnet. This power of attraction and repulsion, in other words the power of love and hate is something wonderful, particularly in an inanimate object like the magnet, and the power of each pole in recognising the different poles of other magnets and behaving always in the right manner suggests superb intelligence. Here south is called negative and the north positive.

The south pole or the negative attracting north or the positive, and the north or the positive attracting south or the negative is a law almost universal in the world.. This is called the law of polarity. About this law of polarity, Dr. Oscar Brunler has written in his remarkable book: *Rays and Radiation Phenomena* (p. 6):

"The law of polarity may be called the first and

fundamental law of all creation."

"Only man and woman together can create a child."

"Only a positive and negative electric current can produce an electric light."

"Only an acid (positive) and an alkali (negative) can produce a chemical compound."

"Only the joining of the two—the positive and the negative—can create a new manifestation."

Dr. Brunler says further:

"Two acids cannot combine and form a chemical compound."

"Two positive currents cannot produce a light."

"Two positive poles of a magnet repel each other."[1]

The north pole of a magnet is positive while the south pole is negative, and the intervening space between the two, north and south, is neutral. Speaking philosophically, the magnet is a miniature universe with all the forces operating in the cosmos. The positive, negative and neutral forces are operating in the universe and these are called the Sattva, Rajas and Tamas in philosophical works. Sattva is positive, Rajas is negative and Tamas is neutral. These are the three cosmic qualities which are deitified in the Hindu pantheon in the forms of Siva, Brahma and Visnu.

The five great elements are also to be found in the magnet. These elements are the earth, water, fire, air and ether. The north pole is constituted of the forces of earth and water, the south pole is fire and the neutral zone consists of air and ether.

---

1. In the highly original work of the late Dr. Oscar Brunler, entitled *Rays and Radiation Phenomena* much information about the law of polarity, Cosmic colours and the bio-cosmic energy is given. Here in this book many of the opinions are coloured by the author's views. It is a brilliant book full of most original suggestions.

It may be noted that according to many systems of philosophy in the world the universe is composed of these five primordial elements, and as such they are also to be found in the magnet.

The five great elements each contribute one of the five forces or Vayus according to Indian scriptures. The five forces are Prana, Apana, Samana, Udana and Vyana. Earth contributes the Prana Vayu, Water the Apana Vayu, Fire the Samana Vayu, Air the Udana Vayu and Ether the Vyana Vayu. The north pole of the magnet contains the forces of Prana and Apana, the south pole bears the Samana Vayu, while the neutral zone contains the Udana and the Vyana Vayus. All these Vayu forces are to be found in the magnet if a little hard thinking is resorted to. Thus the north pole is cold, south pole is hot, while the intervening portion is to be called neutral. In atomic parlance, the north is proton, the south is electron while the middle portion is neutron.

Further reflection will show that the seven planets can also be discovered in the magnet. The seven planets are: Sun, Moon, Mars, Mercury, Jupiter, Venus and Saturn. The sun and mars are hot forces and are negative in character, and they reside in the south pole of the magnet. Mercury, Venus and Moon are cold forces and therefore they are positive in character. They reside in the north pole of the magnet. Jupiter and Saturn are neutral forces and they reside in the space intervening between the north and south poles in a magnet. The seven planets, according to our scriptures, envelop every tangible and intangible manifestation in the universe in the form of seven cosmic rays.

With a little more reflection it will be apparent that the seven cosmic colours or the seven rainbow colours also reside in the magnet. Violet being the colour of Saturn is neutral.

Indigo is the colour of Venus and it is positive. Blue being the colour of Jupitar is neutral. Green being the colour of Mercury is positive. Yellow being the colour of Mars is hot and negative in character. Orange being the colour of Moon is cold and therefore positive. Red being the colour of the Sun is hot and therefore negative in character. The north pole of the magnet being positive is composed of Green, Indigo and Orange, while the South pole being hot is composed of the cosmic colours Yellow, and Red. The neutral portion is composed of Blue and Violet. Thus it can be seen how the seven cosmic colours take their position in the tiny little thing we call magnet.

Thus the magnet is a miniature universe and all the forces that are operating in the universe are seen operating in this little thing also. The manifestation of power inherent in the magnet is divine in character. Its power is derived from the source of all powers in the universe according to the all-powerful cosmic law. In its orignal state of lodestone its magnetic power is manifest, and this is given by no human agency. The magnet is thus endowed with divine power, and being divine it has the three principal divine qualities, of omniscience, omnipotence and omnipresence. Those who have eyes to see will find these powers easily. To those who are blind, the magnet will ever remain as a curiosity and a toy.

The omniscience inherent in the magnet is of supreme importance in magnet dowsing. How this omniscience can be exploited for human good and for divination purpose is the subject matter of the present brochure.

But there is also Omnipotence in a magnet which imparts to it wonderful healing properties. These properties are described in the third chapter of this book which is contributed by my friend and colleague, Mr. Howard D. Stangle of New York at my request. This chapter will reveal some great secrets attached to the magnet not known anywhere hitherto.

The present writer has some experience of the healing power of the magnet, and a few concrete cases are referred to here. A gentleman (45) suffering from chronic and painful arthritis of both the knee joints for nearly ten years could not squat for dinner nor could sit in the closet. Two knee caps were brought and two magnets were introduced into these caps and the patient was asked to sleep with the magnets. This simple treatment made him fit in a month's time, all handicaps disappeared completely and he was able to resume normal work. He valued the magnets and used them for years thereafter.

Another lady (20) came with a painful swelling on the arm just above the wrist. It was two inches in length and quarter inch high, troubling her for several weeks. Instead of giving any medicine a magnet was bandaged on the swollen part, and she was asked not to remove the bandage for two days. On the third day when the bandage was removed, the swelling was gone. Her thanks were profuse, and she returned the magnet.

Mr. Howard D. Stangle the author of the third chapter hereinafter reproduced has been experimenting with the magnet now for many years with varying results. Today he may be considered as the highest authority on the subject of magnet healing.

CHAPTER III

Its Healing Power

(By Mr. Howard D. Stangle of New York)

The contributor of these paragraphs believes Magnetism to be a true science. It is all along the line admitted that the magnet has hidden properties, and that something of these has from time to time become known, made use of, and even held in reverence. Personally, however, we are in no position to lay claim to this knowledge. One may grasp the fact of magnetism being a force everywhere present; its place is in everything animate and inanimate; thus like light and life the source of terrestrial and human magnetism is in the sun. This much is self-evident. Magnetism is the *sakti*, the passive condition of electricity. There being a "sun" in everything, at the centre of everything, whether of man, plant, animal or precious stone, it is not difficult to see the connection with the term "life" as generally understood, and how the magnetic fluid may be actively transmitted, as in the case of "animal magnetism", or latent, as in the mineral.

In some Hindu works the spleen is said to be the principal organ of transmission of vital life and energy. We have read somewhere that the spleen radiates its power in curves not unlike those patterns given to iron filings by the steel magnet. Here then would seem to be a link of comparison between the metal magnet and that all-important organ of the body, the spleen. Were more known today, of what has been forgotten in the past, it is possible that we should have a clearer view of the entire human organism constituting an electrical and magnetic field or—a magnet of a superlative kind.

It seems that Magnetism is a subject that should excite universal interest. A Frenchman of the last century, Dr. Francis Victor Broussais made the statement: "If magnetism were true, medicine would be an absurdity." Our position is that magnetism is among the truest things in life on earth, and seems to be shared by our most eminent men of science—nor merely of the West, but around the globe. We need offer no apology then for quoting from that scientist who proved his rank among the greatest—Paracelsus, who in speaking of the properties of the iron magnet, said:

"........there are qualities in a magnet not known to every ignoramus, and one of these qualities is that the magnet also attracts all martial humours that are in the human system........Martial diseases are such as are caused by auras coming and expanding from a centre outwards, and at the same time holding on to their centres; in other words, such as originate from a certain place, and extend their influence without leaving the place from which they originate. In such cases the magnet should be laid upon the centre, and it will then attract the diseased aura towards the centre, and circumscribe and localize the disease, until the latter may be re-absorbed into its centre. It is useless to try to suppress the external symptoms that are caused by a disease, if we at the same time allow the disease to spread. A poisonous tree cannot be kept from growing if we simply cut off some of its branches or leaves, but if we can cause the vital essence which it draws up by its roots from the earth to descend again into the roots and re-enter the earth, the poisonous tree will die on its own account. By the attractive power of a magnet acting upon the diseased aura of the blood in an affected part, that aura may be made to return into the centre from which it originated, and be absorbed therein; and thereby we may destroy the herd of the virus and cure the patient, and we need not wait idly to see what Nature will do. The magnet therefore is especially useful in all inflammations, in fluxes and ulcerations, in diseases of the bowels and uterus, in internal as well as in external disease"………(N.B. "If we remember that the blood corpuscles, and consequently also the nerve aurae, contain iron this statement appears very rational").

It is to be noted that the above treatment is accorded by Paracelsus to those types called "martial" (Mars ?) diseases; and that in cther instances the negative or repulsive pole of the magnet is employed. "It should, however, be remembered that the manner of using a magnet changes according as to whether we wish to draw the diseased aura out of the body, or to cause it to be reabsorbed into its centre". And again, "There are great many diseases that may be cured by the proper use of the magnet, but for those who are able to understand such things the hints already given will be sufficient, while those who have little understanding would not comprehend this system even if we are to write a book about it".

In studies of this unfamiliar, or rather uncharted nature, requiring of necessity some individual pioneering, it is natural to make first experiments upon oneself. One fact brought home in the process is the old truth that Nature proceeds cyclically, that Nature is a slow plodder and will demand her *own* time. We find that impatience of seeing results will avail not at all. Perhaps this is best, as the investigator must perforce also take time to make his researches in a more thorogoing manner than is his custom. But as one learns, one profits.

May we now beg the reader's indulgence in a modest account of a few personal case histories. A man in his late 'sixties suffered periodical recurrences of the fruits of an enlarged prostrate gland. During these flare-ups which sometimes lasted for several weeks without intermission, the pain was intense, all business activities were suspended, and there was little rest and sleep. Only surgery was prescribed to bring relief, and this the man wisely declined as there was no cancerous state indicated. We one day approached him with the suggestion of wearing a magnet, and to our great amazement met not the slightest opposition to the proposal. This man did not even raise questions at the time—but we may add that during the years that have since intervened he has become a devotee to the study of the magnet. Some few months passed, when one day the patient abruptly said: "Well, I'm still wearing the magnet !" We thanked him for the information, being quite glad to learn that not only had there been no untoward results from continued use of the magnet itself, but that there had been no hint of pain in that area. We may as well confide the sequel in this case, which is that the man still wears the same magnet, and this at a distance of several years from the day he first put it on, and has never for one day relapsed into the old condition, nor lost a day from his illness.

Our method of applying magnet is directly to the skin at the disease centre, or within the area of the centre ; and we have found that diseases in the main require the positive, or "attractive" pole of the magnet in order to cause the disease-aura to retreat back into that "seed" or centre whence it sprouts and spreads. We believe Dr. Bhattacharyya may fully elucidate this theory in connection with the action of the Vital Airs, Fires, and Waters, which all have their separate and permeative loci in the body, which loci or areas might thus provide the hypothesis of why degenerative diseases may spread from organ to organ, as in some forms of cancer, for example.

Another case coming under our observation was that of a man in late "fifties, who had through the years suffered periodical attacks of violent nature, seemingly pointing to the "heart condition" today so common in the West. When we made our several diagnoses we were troubled by the fact that the heart seemed sound and that the centre of his illness had to be sought elsewhere. This of course was contrary to other diagnoses and to the patient's own opinion ; yet we found the heart to be normal and strong in our own tests and proceeded upon that ground. We evolved the theory as a working basis that a common cause underlay his frequent attacks over a long time which were thought variously to be independent diseases. We will not name the diseases, as the theory would seem a bit farfetched even to the interested student ; suffice to say that this working basis was formed on the idea of incipient cancer, or what might some day take on that form, in the muscles of the lumbar area. Accordingly we persuaded the patient finally to begin wearing the positive side of magnet next the bare skin—of the hip, at a sore spot. The patient wore it good-humoredly, though he laughed at what I affirmed as the then locus of the disease-aura.

Now whether right or wrong in the deduction, the patient said a few days later that already relief had set in for the first time during that episode of recurrence ; had, in fact, set in about two hours after the magnet was taped to spot on hip The magnet was thus worn, in this case several hours daily, for a matter of several months. During this period the "heart condition" gradually disappeared ; the pain lessened ; finally there was no longer need to avoid stairs ; digestion righted itself with other matters of the economy, and so on. This man finally agreed there could have been no heart trouble as diagnosed by his doctors. The heart had simply a double load to carry, which was brought on by the sapping of the vital processes consequent upon the unnatural presence of such dweller (cancer) within the organic and magnetic fields.

CHAPTER IV

Dowsing Technique

Radi-esthesia is the science dealing with the study of energy fields by sensitive individuals or the study of the human body's perception of, and reaction to ultra-low intensity radiation. The operation technique is known by various terms such as, Dowsing, Divining, etc. which mean the same thing. Radi-esthesia mostly deals with the detection of disease in human beings whereas Dowsing deals with the detection of underground water, oil, minerals etc. There is a rule for everything, for sitting, for direction, for the hands and for the feet, and for the magnet, the charts and the phials of medicine required for the purpose of dowsing. All these constitute the discipline of the operator and the more he is able to master the rules, the more successful he becomes. Slipshod, careless and inattentive work is not rewarded with good results in magnet dowsing. Needless to say, dowsing should not be attempted when one is tired physically and mentally. The mind should be calm and composed.

First of all, the operator has to face one direction, namely West if he is right handed. But if the operator by any chance is left handed, then he should face East in order to get results from magnet dowsing. The operator should sit on a chair and a small table should be in front of him where charts, phials, photographs, etc. should be kept for the purpose of examination.

The feet should be placed firmly on the ground and the

legs should not touch each other or cross each other. The right and left legs have their own polarity, and they have their own positive and negative character, and if by any chance they touch or cross each other tests will be vitiated.

The right hand should hold the pendule with two fingers only, namely the thumb and the first finger. These two fingers should not come in contact with other fingers nor should touch them. Good care should be taken of this because if other fingers touch the two holding the pendule, all tests will be vitiated. It may be noted here that every finger of the hand has its own polarity and if the unwanted fingers touch the thumb and the first finger with which the pendule is held it is quite reasonable that the tests will be invalidated. There is always a right and wrong way of doing things, and here in magnet dowsing the wrong way should be avoided.

The left hand should hold the magnet to magnetise the body and mind of the operator. This magnet should be of moderate strength. For all practical purposes a pocket horse-shoe magnet is found to be suitable for dowsing purposes. Care should also be taken that the two hands do not touch or cross each other. The left hand holding the magnet may be raised and kept at a distance from the right hand holding the pendule, or it may remain hanging on the left side of the body. If the left hand magnet remains too near the right hand or the chart on the table, it may have adverse action on the tests.

The table on which the charts or medicines are placed for the purpose of dowsing, should be clean and should not have any other articles thereon because otherwise the tests may lose their precision. Nothing extra should be on the table because everything has its own vibrations and these may vitiate the tests undertaken by the operator.

Before sitting for work, it is advisable to wash the face, hands and feet with water. This will keep the operator clean and his mind will be at ease, and these would help him in dowsing correctly.

There are many days which should be avoided for the purpose of dowsing because these days are inharmonious for testing purposes. These days are usually connected with the sun and the moon, and when their magnetisms are at variance magnetic influences on earth are also at fault. Therefore, it is desirable to avoid certain days for testing purposes. These are the Full Moon and New Moon days, the 1st, 8th and 14th day of the Moon for every fortnight. Thus four days only in a fortnight should be regarded as prohibited days for dowsing.

The time when the sun is rising or setting is not good for testing, and therefore sunrise and sunset time should also be avoided. These are the few rules to be remembered by the operator and if he follows them he is likely to get quick mastery over dowsing.

Before starting dowsing the first question that should always be asked is "Whether it is the proper time for dowsing or not". If the pendule moves up and down then one can proceed further with the work, but if the pendule moves horizontally then work should be suspended for 15 minutes before starting again.

All the questions asked should be asked separately as definite statements. For example, a statement may be made such as: 'This person requires more than one colour'. If the pendule moves up and down, the statement is correct. But, if it moves horizontally, then the statement is false, and the person requires only one colour. If the statement is found to be correct, it means that the patient has a colour hunger for more than one colour, hence, the next statement should be that 'He requires more than 2 colours', and so on till the pendule says No. After knowing the exact number of colours required by the person, the first colour should be found out, then the second and so on.

Now the above may be illustrated by means of a few drawings in order to show the correct position of the hands and the pendule, and the manner of dowsing with the help of charts and medicine phials, just to help the operator for the work of dowsing. This chapter gives a small, universal and easy chart for practice purposes and to get preliminary answers for questions asked to the pendule. The questions should be written on a piece of paper and placed on the centre of the base line. Over this question the pendule should be held in the right hand, and the left hand should carry another magnet. When a gentle rotatory motion is given to the magnet pointer in the right hand, and then the pendule is allowed to move by its own force, it will give the answer by being attracted either to the right or to the left. If the magnet goes right the answer is Yes, but if it goes left the answer is No. This rule applies equally to all kinds of dowsing tests.

Here, it may be mentioned that different dowsers adopt different techniques to interpret the movement of the pendule. Most of the Western dowsers interpret the clockwise gyrations of the pendule as Yes or having attraction for the item tested and the anticlockwise gyration as meaning No or having repulsion. These various movements may be explained in the following way. When two people converse with each other they converse in a language which is understood by both.

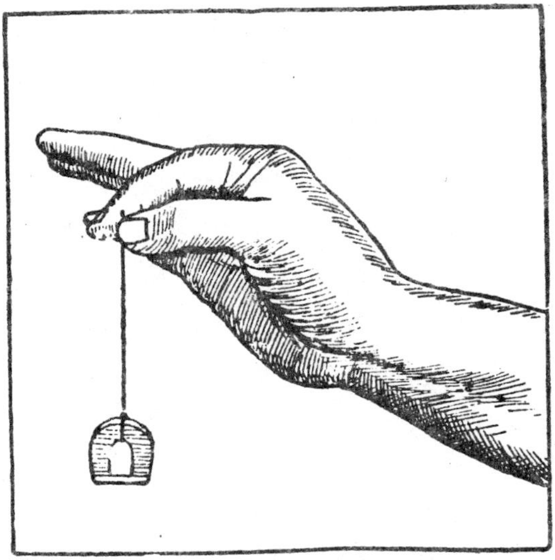

Fig. 1. Magnet Pointer

Similarly, dowsing means that there is a conversation between the dowser and the pendule. An agreement is made with the pendule about the interpretation of its movements. Hence, if a dowser has an agreement with a pendule that when it moves in a clockwise gyration it would mean Yes, then he will always have the two movements clockwise or anti-clockwise. Whereas, another dowser who fixes the up and down movement as meaning Yes and the horizontal movement as meaning No, will always get these two movements from the pendule. Dowsers may use the methods which prove most successful to them personally. In this book the author has fixed the two movements namely "Up and Down" meaning Yes, and "Horizontal" or parallel movement meaning No.

The Magnet Pointer is shown in Fig. 1. The picture also shows how the pointer should be held in the right hand

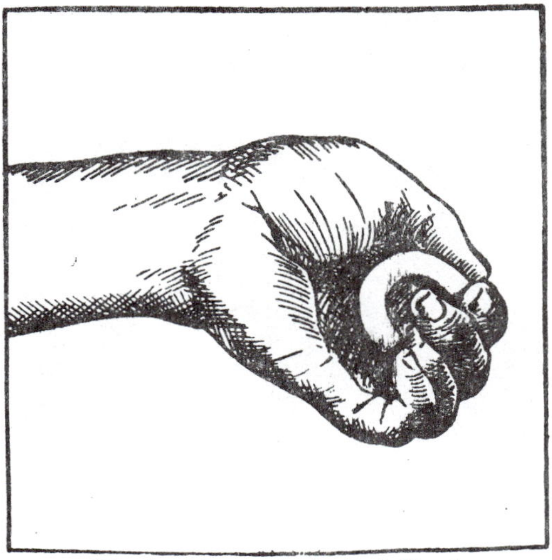

Fig. 2. Magnet in the Left Hand

with the two fingers, namely the thumb and the first finger. It is to be noted that the two fingers holding the pendule do not touch any other finger of the right hand. The magnet is shown here as suspended by means of a black string about 3" inches long. The string is tied to the centre of the bent side of the horse-shoe magnet. This pendule should be held over charts and photographs, and operated in the manner previously indicated.

The horse-shoe magnet required for magnetising the operator's body and mind is shown in Fig. 2 which is held in the left hand. The magnet should always be held with the keeper removed from it. The left hand may be held aloft at a distance from the right hand pendule, or it may be kept hanging on the left side. This magnetisation is very important in all kinds of magnet dowsing, whether crude or

Fig. 3. Dowsing with Chart

advanced, because if the body and mind are not magnetised the results arrived at may be wrong and undependable.

The actual dowsing process is shown in Fig. 3. Here, the Percentage chart is being used for the purpose of dowsing. On this Percentage chart, in the centre of the base line a photograph is placed for the purpose of testing. This photograph represents the person under examination. The magnet pointer is held in the right hand while the hand magnet magnetises the mind and body of the operator. Where the photograph is not available for dowsing purposes, any substitute may be placed, such as blood smear, sputum specimen, signature, thumb impression or a name card of the person in question. In the higher cosmic sphere there is no difference between the Name and Form, and any substitute of a person can be recognised as belonging to him by the cosmic forces.

The photograph on the base line may be tested in two ways: firstly, it may be magnetised by placing a magnet on the photograph itself or its many substitutes referred to above. Secondly, if the photograph is not magnetised then a magnetic pointer as shown in Fig. 1 should be used for the purpose of dowsing. Generally, the rule should be to magnetise the photograph if the chart has many entries, but if there are only a few entries on the chart it is not necessary to magnetise the photograph because the magnet pointer itself will be able to do it. Where there are many entries on the chart, a pendule with a pointed end will have to be used. It may be of wood, metal, bead or silver. When the testing is done on the Electro-magnet Box, there is no need to place a magnet on the photograph or any substitute of the person, as the field is already highly magnetised.

When everything is ready, the right hand pendule should be given a rotatory movement from the left to the right. The pendule will begin moving, and then it will move by its own force, and soon develop a straight motion from its elliptical movement, and point to one or another of the many entries on the chart. If the percentage is 75 or less, there is no danger to life. If it is between 90% and 95%, it shows that it is a serious case, but the danger to life is not very great, but if it be 95% to 100%, it should be understood that the life currents are getting exhausted and great caution should be taken to treat the case. This method is common to all forms of dowsing, and therefore deserves to be noted with care.

When it is a question of selection of medicine from among many, the method of dowsing should be as shown in Fig. 4. Here on the table a number of medicine phials are placed, arranged in a semicircle equidistant from the centre where

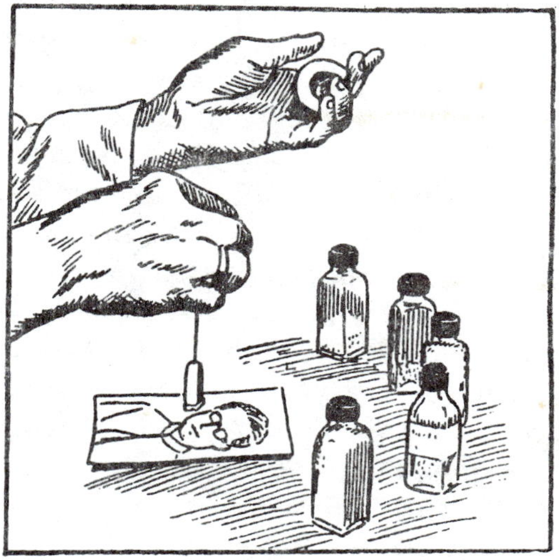

Fig. 4. Dowsing Medicinc

the pendule is held and operated. The left hand carries the magnet to magnetise the body and mind of the operator. The magnet pointer is suspended from the right hand over the photograph of the patient placed on the centre of the semi-circle or rather its base line. The stage is now set for dowsing the correct medicine with which the patient is in harmony.

With the right hand a circular and gentle rotatory movement is given to the magnet pointer and soon it begins to move by its own motion. As it moves slowly from one point to another, the movement of the pendule should be watched with care. Later the movement of the magnet becomes constant and points to one or another of the five bottles placed before it in a semi-circle. The experiment should be repeated at least ten times before a decision can be taken and the right medicine is fixed for that particular person whose photograph is placed on the chart.

This simple experiment will be found useful for all who have anything to do with healing. That medicine which has cosmic harmony with the patient in question is the best medicine for the time being. It should however be remembered that the pendule will show the strongest and the most appropriate medicine amongst those placed before it, but it cannot obviously point to the medicine most suitable and curative unless it is placed before it. It is because of this that some dowsers fail to get the very right medicine and are often disappointed. The strength of the selected medicine may be further ascertained by the Percentage Chart. If the selected medicine shows 75% then it should be concluded that it is not the curative medicine, search for which therefore should be continued. The indicated medicine must show 100% before it can be administered. These are some of the rules to be remembered for the right selection of medicine by dowsing.

Another method which may be adopted by the busy practitioner is quite simple. A chart should be prepared with the 26 alphabets and on it dowsing should be done to find out the alphabet of the medicine needed by the person. Once that is found, all the medicines beginning with that alphabet in the Materia Medica should be tested and then the correct remedy will be found. On finding the medicine it should be double checked on the Percentage Chart. If it shows 100%, then it can be given with confidence. By this one simple method many of the so-called incorrect prescriptions may be avoided to the eternal benefit of the suffering and ignorant public.

With these preliminaries a chart for practice purposes is given in Fig. 5 which may be made use of rather extensively

Fig. 5. Chart for Practice

for various questions arising in the mind of the operator. This is a simple chart with two entries namely: YES and No. in the right and left respectively. Although simple, it can yield valuable knowledge of a super-sensory character. In life we are always faced with two alternatives, and we require guidance always, however defective that guidance may be. Whether a patient will live or die every doctor wants to know. The magnet can answer this question. Whether a candidate will pass his examination or not can be answered by the magnet. Whether a person will be convicted of an offence or will be acquitted, will be answered by the magnet. An answer coming from an inanimate object like the magnet is certainly better than the conclusions of an astrologer or the crystal gazer. The wise man should take advantage of the opportunity offered to him.

CHAPTER V

Magnet Study of the Individual

The study of the individual by the dowsing process is of primary importance to the operator or the physician who is required to handle a large number of individuals. When a patient comes, it is necessary, for instance, to know whether he is short-lived, or possesses a middling or long life. It is also desirable to ascertain his brain power in order to find out the group to which he belongs. The physician also should know the true colour hunger from which the individual is suffering, what gems or jewels will be beneficial to him, what planets are afflicting him, and also the manner in which he is likely to meet with his death. These and other endless questions of importance can be answered by magnet dowsing with the help of the charts illustrated and described in this chapter. Any information about the individual obtained by dowsing will be of value to the physician who is called upon to handle him.

Now-a-days tall claims are being made for the so-called lengthening the life of every man by medical science. According to it every man can live indefinitely provided that he is careful enough to get himself examined every now and then and take the necessary drugs and injections. The present writer has scant respect for such claims, because as we see everywhere even the doctors themselves die very early and many rich clients who are constantly taking medicines and injections are seen to die even before they are

Whether a gem is genuine or fake can be answered by the magnet. The gem should be placed on the centre of the base line of the practice chart, and the pendule should be operated over it in the manner previously set forth. If the pendule goes to Yes it is genuine, but if it goes to No it is fake.

Tests should be repeated at least ten times, if possible, before a decision is taken. For every test with the magnet this is essential and should not be neglected under any circumstance.

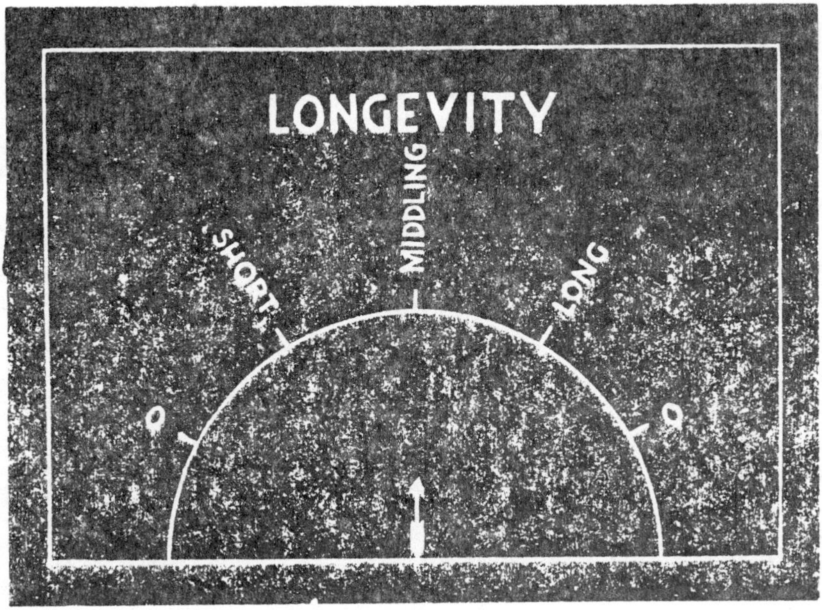

Fig. 6. Longevity Chart

fifty. Even big political leaders and heads of States are no exception to the rule.

But the old idea is, that as soon as a person is born, his life period is fixed and it is possible to divine his longevity through planetary positions at birth. According to this, some are short-lived, while others have either medium or long life. By magnet dowsing this very thing can be detected with the help of the chart illustrated in Fig. 6.

The photograph or the name card of a person should be placed on the centre of the base line and by operating the pendule in the manner previously indicated, precise longevity should be divined. If the pendule points to "short" then the maximum life period is 33, if it is "middling" then it is 64, but if it goes to "long" the life-period is 84 and beyond. This longevity should be ascertained in every case before any action can be taken to relieve the individual's distress. The experiment should be repeated several times before a decision can be taken. A little practice will enable the operator to dowse longevity properly. Once we know the life term it is possible to regulate our actions. The ages mentioned above are with reference to men. For animals and others dowsing should be done with the help of the Year Radiation Chart in order to find out the last year of its life. I again caution the operators to keep the knowledge a secret as the information obtained is only for the use of the operator to know the gravity of the case in hand.

To ascertain the brain power of an individual the Biometric Chart illustrated in Fig. 7 will be found useful. These radiations are almost fixed from birth and it is hardly possible to make great changes in the figures of brain radiation arrived at by using this Biometric Chart. It helps to select children for higher studies, as a child with a low biometric reading should not be pushed for higher education as the efforts of the parents will be completely wasted.

Different brain radiations sharply divide several classes of people. The brain radiation, when high, shows high power of understanding of human problems with superior human sympathies. Low vibrations likewise show lower understanding and lack of sympathy for the suffering of humanity in general. The religious reformers show very high radiations such as 700 on the chart and they have the greatest understanding and sympathy for human sufferings. Between 600 and 700 are the people who may be called seers which include

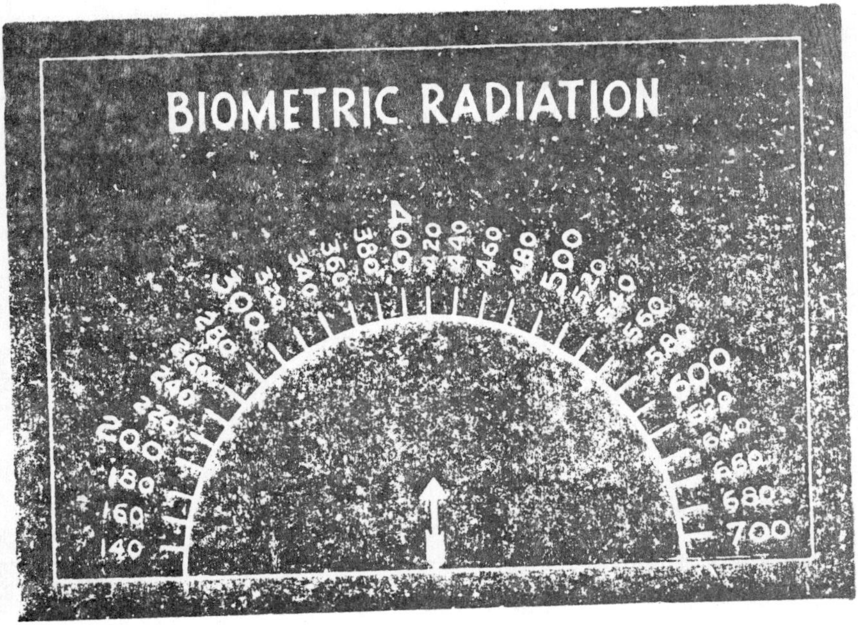

Fig. 7. Biometric Radiation

great poets, dramatists, and great men of science who have produced immortal works and benefited humanity at large for a long time in history. Between 500 and 600 are persons who are endowed with great originality, and include, among others, original workers in the field of literature, philosophy, science and arts, in music, sculpture and painting. According to Mr. Brunler actors are not able to hold the audience if their radiation is below 500.

Between 400 and 500 are persons having the utmost respect for the printed word which they take as gospel truth. These include professors, famous teachers, text book writers who are credited with phenomenal memory and they can

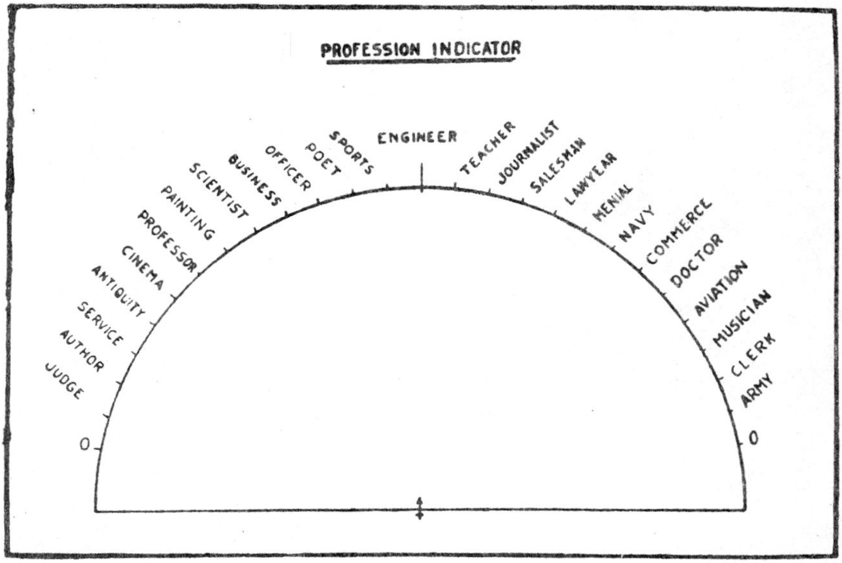

Fig. 8. Profession Indicator

repeat Shakespeare, Milton and Kalidasa almost non-stop like so many gramophone records. These persons are conscious of their importance but are bereft of originality. It is very risky to discuss original matters with such persons, particularly if such matters are not yet out in print.

School teachers, ordinary practitioners and pleaders of the plodding type come under Biometric radiations of 300 to 400. Maid-servants, peons, errand boys, and the rest generally radiate between 200 and 250 Biometric. Below 200 are usually the people with very low brain power and they include idiots, morons, and imbeciles.

The Profession Indicator Chart (Fig. 8) may be consulted to find out the suitable profession indicated for a child. There are several entries in this chart and dowsing should be done very carefully before coming to a conclusion. This will help parents to coach their children for the particular profession indicated for the child.

Both living and dead people give brain power radiations and these can be measured with the help of the chart now under discussion. When a person is living, his photograph can be used, but if he is dead, his name card should be prepared by the operator himself and placed on the chart for examination by magnet dowsing. By this process, brain radiations of every person can be ascertained. To my mind, this test is very important, and by this test it can even be ascertained whether a boy or girl is fit for higher university education, and whether a particular person, because of his low brain radiation, is fit to occupy the post of a professor By this Biometric test many fake people in every department of life can be found out.

Books also have their radiations, and it can be tested by the dowsing process. Instead of the photograph the book itself or its name card may be placed on the centre of the chart, and the pendule when operated over the book will indicate its strength. Books having more than 600 Biometric are worth serious study, those with 500 to 600 Biometric have some permanent place in literature and can be read with interest, while those below 500 are ephemeral and are not likely to survive long because they lack permanent interest.

Of all subjects the subject of colour is of supreme importance for every one.[1] These are the seven VIBGYOR

---

1. Dr. Roland Hunt in his epoch-making book: THE SEVEN KEYS TO COLOUR HEALING, published by the C. W. Daniel Company Ltd., Essex, England has given authoritative indication of how colour is important for our lives, and the different methods by which colour vibrations can be imbibed for our physical, mental and moral health. I have learnt much from this book.

colours along with the two invisible colours, Ultra-Violet and Infra-Red. The first seven cosmic colours are seen on the Rainbow with open eyes, and they are seen everywhere moving in a different dimension which is revealed by the prism. All tangible things in the world are nothing but condensations of the seven cosmic colours and rays and radiations.

Old Indian scriptures like the Kurma Purana affirm that the body of the Lord the Creator of the Universe which is the same as the universe itself, is composed of cosmic colours and among them seven are the chief being the matrices of the seven planets. These cosmic colours are endowed with the three divine qualities of omniscience, omnipotence and omnipresence, and with the help of these qualities they create, maintain and destroy the universe. In fact, according to our ancient scriptures the cosmic rays are the lieutenants of the Primordial Cosmic Consciousness by whose thought the universe came into being.

Thus, the cosmic colours are at the root of everything tangible, and the Five Great Elements: Earth, Water, Fire, Air and Ether are to be found among the seven cosmic colours. The element of Earth is Green, water is Orange and Indigo, Fire is Yellow and Red, Air is Violet, and Ether is Blue, and thus the elements are distributed among the seven cosmic colours. The colours are the subtle forms of the five great elements.

We are continuously receiving from the cosmos our share of the seven cosmic colours which are nourishing our body, mind and our individual soul. Whenever one or another of the seven cosmic colours becomes deficient we suffer from disease. The disease with its painful symptoms is nothing but the cry of the individual for the right cosmic colour which is deficient. Disease may therefore be taken as nothing but the hunger for cosmic colour. When that colour hunger is satisfied by the supply of the true colour disease disappears. The deficient colour required by the individual can be supplied in many ways. By giving medicine of any system of medicine that colour hunger can be satisfied, but it is not possible for us to know the true colour of the medicine. The same medicine may be given in another case but it fails because the colour hunger is different from the previous one. The colour can be supplied by Teletherapy which deals with invisible transmission of colour vibrations over photographs of respective patients. The colour hunger may also be satisfied with gem medicines containing the principal cosmic colours. Those who are interested in Gem Therapy are advised to refer to my earlier book entitled, Gem Therapy (2nd. Edition) published by Firma K. L. Mukhopadhyay of Calcutta.

When a person is suffering from high fever, the presumption is that it is caused by heat or the element of fire, and therefore, cooling medicines of the element of water are given to cure the person of his fever. Fever is thus a colour hunger for cooling colours: Orange, Indigo and Green. Out of the three Green has the highest cooling effect. If over the person's photograph Green colour is radiated by Teletherapy the person will be cured in no time. Whichever medicine or food has the strongest element of Green in it should cure the fever without doubt. Especially when the fever relates to the element of earth, such as measles, small-pox, chicken-pox, typhus or typhoid, Green is doubly indicated and it cures easily, gently and without medicine, by the vibrations of Green cosmic colour alone. Even an Emerald placed in water overnight will impart its Green colour to the same. This water charged with Green may be given to dring in anticipation of good result.[1]

Every medicine, as everything else in creation, contains a principal colour. Only in rare cases complex colours are found. It is necessary for every one to know this colour element in the medicine before his knowledge of the medicine may be complete. But unfortunately, no effort has been made to discover the element of cosmic colour that is inherent in the thousands of medicines we use. No wonder, our knowledge being incomplete, we meet with dismal failures while treating cases for a cure. Every Homoeopathic medicine has a colour and this has to be ascertained before its true character can be known or can be prescribed with precision and efficiency.

Let us illustrate this with an example. Capsicum (Cayenne pepper) for instance is long pepper potentized, but the original cosmic colour of pepper is Red because it is pungent. Therefore, when we prescribe Capsicum in potencies it is imperative for us to know that we are introducing the Red cosmic colour or the element of heat into the system of the patient for a cure. The metal Copper, for instance is of Red cosmic colour, and when we prescribe Cuprum Met in potency it will be good for us to know that we are introducing cosmic heat into the patient for a cure: the difference consists in the fact that while Capsicum is short acting, Cuprum is deeper, much deeper in action and long lasting. The heat of Cuprum is required in collapse, when animal heat is wasted away, for instance, in cholera and other quickly killing diseases.

1. Dr. Guyon Richards in his brilliant book: THE CHAIN OF LIFE, second edition, published by the Health Science Press, Sussex, England, besides giving various subjects of supreme importance to every man living, has referred to the all-important subject of colour and its application for human good in Chapter XIX pp. 188ff. All who are interested in colour therapy should never miss his brilliant suggestions.

Another metal, Silver, is an Indigo cosmic colour concentrate, and it has a remarkable cooling effect. In all chronic Pitta (heat) cases with bad liver and spleen Silver can be prescribed with confidence fully knowing that we are introducing cosmic cold into the system.

Colour is the most important factor in life. Our education will not be complete unless we have a thorough knowledge of cosmic colours which create, maintain and destroy everything in creation, everything tangible in the universe. Endless investigations can be made with reference to cosmic colour, and those who will take to these colour studies seriously will be held as the wisest men of tomorrow.

The colour radiation chart (Fig. 9) is therefore important, because it helps us in ascertaining the element of colour inherent in everything tangible in the universe, including men, Medicines and maladies. While testing an object it should be placed on the centre of the base line of the chart, and the pendule should be operated in the manner indicated previously. By these tests the required cosmic colour inherent in everything tangible can be found. Where the object is not available for test, its name card should be placed on the chart. In the Electro-magnet Box the colours are indicated by the bulbs set there and hence this chart may be eliminated.

Experiments in which the cosmic colour inherent in un known medicines, is involved, should be made on Magnet Boxes or by the special Rule Method advocated in this book and described in the chapter on Advanced Dowsing. These are delicate tests with far-reaching consequences and therefore, should be done by experts and advanced students with the greatest care.

Gems are allied to colour, and although their number is

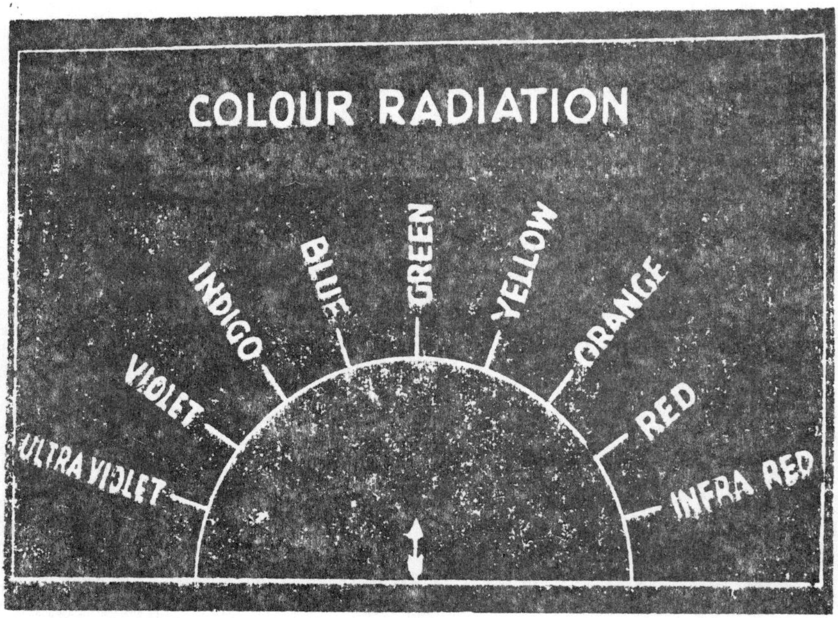

Fig. 9. Colour Radiation

legion, nine among them are the chief because they are affiliated to the nine planets. All gems are not suitable for everybody. Gems are to be selected by a reference to horoscopes, and the evil effects of planets may be warded off by wearing their sacred gems. In order to find out gems suitable for individuals the chart illustrated in Fig. 10 may be consulted with advantage. These gems also have their medical values and they can be used as medicine in various ways. The selected gem may be placed in water and kept there for the whole night. The water charged with the power of the gem can be given to the patient to drink, and it will act as a powerful medicine if the gem is selected with care.

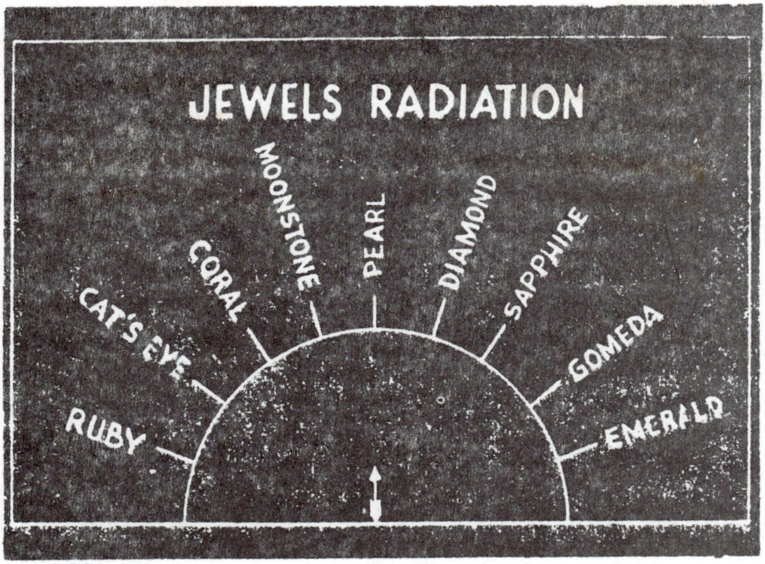

Fig. 10. Jewels Radiation

There are other ways of taking gem medicines and those who are interested in the subject, may consult my book entitled, Gem Therapy (2nd. Edition) published by Firma K. L. Mukhopadhyay of Calcutta.

These gems are again connected with the seven colours of the Rainbow or the seven cosmic colours. Through the gems it is also possible to find out the harmonious and inharmonious cosmic colours for each individual. The gems and their cosmic colours are given below in tabular form thus :

Gems	*Cosmic Colours*
Ruby	Red
Pearl	Orange
Coral	Yellow
Emerald	Green

Gems	*Cosmic colours*
Moonstone or Topaz	Blue
Diamond or Vaikranta	Indigo
Sapphire	Violet
Gomeda (Sardonyx)	Ultra-Violet
Cat's Eye	Infra-Red

On the centre of the chart should be placed the photograph or name card of a person under examination. Over this the pendule should be operated in the manner previously indicated, and the pendule will correctly indicate the gem required for the individual. The information obtained by this process should be confirmed by running the pendule over the colour radiation chart previously described. With the Electro-magnet Box, the colour will be indicated by the pendule moving towards a particular bulb representing a particular colour. As usual, the tests should be repeated several times before a decision is taken.

Allied to the subject of cosmic colour and gem are the nine planets including the two Nodes of the Moon, the Ascending node and the Descending node. Readers interested in Astrology will find the chart illustrated in Fig. 11 as of unusual interest. The cosmic colour hunger developed by a particular patient can be ascertained by a reference to this chart. The planets are connected with the seven Rainbow colours, and if the pendule points out a planet, its colour can be easily known. This planet radiation chart should be frequently referred to, when by other means the cosmic colour hunger has been ascertained.

The colours of the seven planets included in the chart under discussion may be given below in a tabular form :

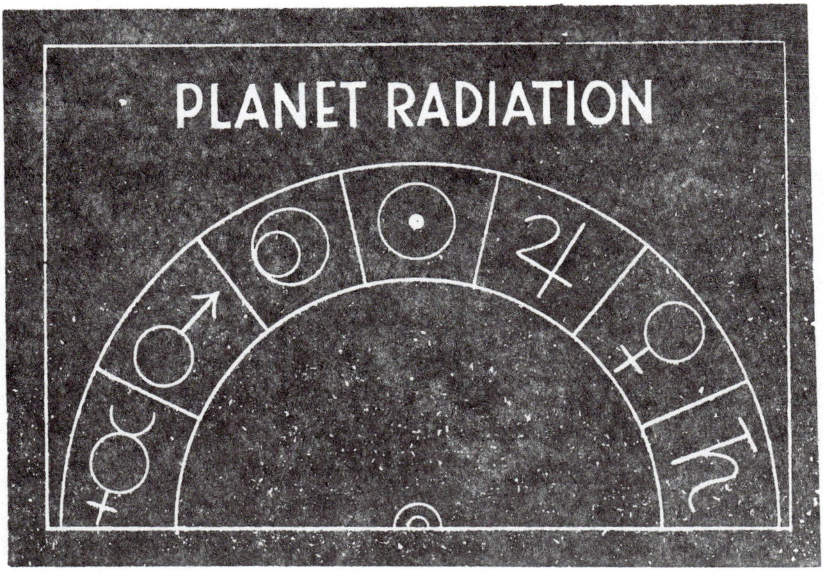

Fig. 11. Planet Radiation

Planets	*Cosmic Colours*
Mercury	Green
Mars	Yellow
Moon	Orange
Sun	Red
Jupiter	Blue
Venus	Indigo
Saturn	Violet

The planets on the chart have been arranged in the above order from the left to the right, and their cosmic colours are also given here in the same order. It may be noted in the Planets chart that the two nodes of the Moon have been omitted. Planets that are malefic should be propitiated by

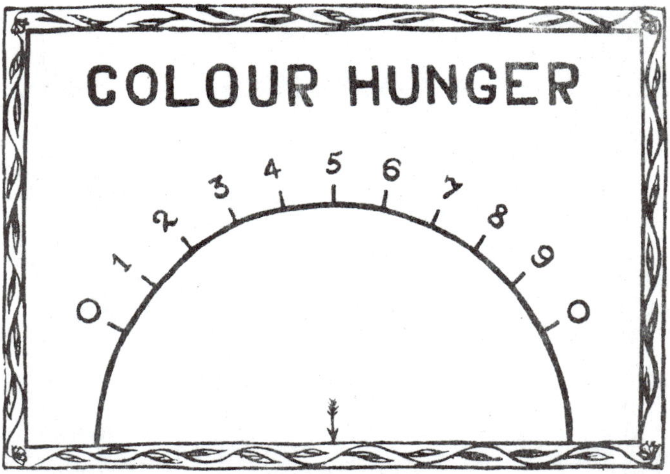

Fig. 12. Colour Hunger

wearing their sacred gems or by gem radiations either by teletherapy or gem therapy.

While ascertaining the planet that is malefic the photograph of the individual may be placed on the centre of the base line and the pendule should be operated as before. The planet pointed out by the pendule is the planet which is not giving protection, that is to say, is not supplying the necessary volume of cosmic colour to the person in question. The colour thus arrived at is required to be regularly supplemented by cosmic colour.

It is not always a fact that an individual has the colour hunger of only one particular colour. Several planets may be malefic at one and the same time so that the person does not receive the protective vibrations or colours from those planets. To find out the colour hunger for a particular person the chart for Colour Hunger (Fig. 12) should be consulted and when the

Fig. 13. Manner of Death

number of colours is found, the colours involved are to be found out one by one by dowsing.

Another important factor about the individual is the manner in which he is likely to meet with his death. We find in this wide world that people die in various ways. Old people thought that the manner of death depended on destiny. However, when it becomes necessary to know the manner in which a particular person would meet with his death, the chart illustrated in Fig. 13 should prove of some value. For divining the manner of death this chart is devised and it has a number of entries which cannot be said to be exhaustive.

To ascertain the manner of death the photograph of the person under test should be placed on the centre of the base line and a pendule should be operated over it as before. The manner of death will be pointed out by the pendule and the tests should be repeated several times before a decision is taken. The information obtained by dowsing in this manner should never be communicated to any one. This test is a very delicate one and all care should be taken to arrive at a decision. If possible, this test should be done on the Magnet Box with a pendule which has a pointed lower end, in order that the movement of the pendule may be watched accurately.

CHAPTER VI

Magnet Study of Time

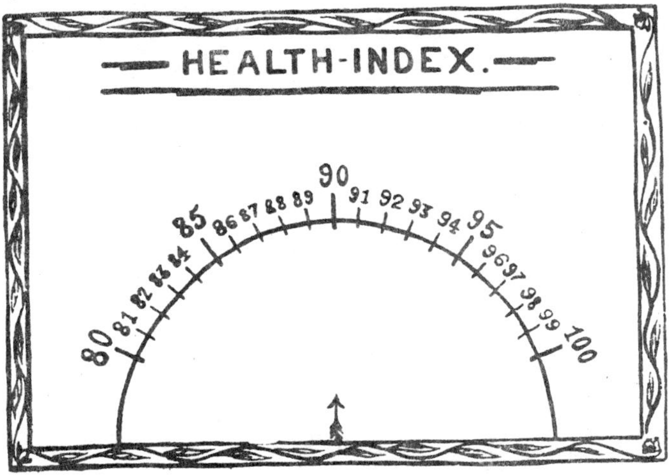

Fig. 14. Health Index

Time is an important factor in life. Time has three phases, past, present and future. All these phases of time are of primary importance to every one and to all forms of life. People are not very anxious to know about the past or the present, but in future all are immensely interested. This all-important Time factor can be studied with the help of the few charts illustrated in this chapter with their appropriate notes and explanations.

One of the charts, the Health Index Chart 80 to 100 (Fig. 14), should always be consulted whenever a person comes

Fig. 15. Health Index

to consult a physician. The photograph or the name card of the person at the centre of the base line with the pointer pendule moving in the right hand will show one of the many entries on the Health Index chart. If it shows 80 to 90, no danger to life can be expected. But, if it shows 90 to 95, then the case is serious but even then it is amenable to proper treatment. When it is between 95 and 100, then there is grave danger to the life and great care should be taken to treat it. When the reading appears to go above 95, the second Health Index Chart 97 to 100 (Fig. 15) should be carefully consulted. Mostly, cases showing 98 to 100 prove fatal and hence regular testing should be done to find out whether the patient is improving or not. On the Percentage Chart (Fig. 16) it should be noticed what per cent improvement is made by a patient after the commencement of the treatment. This always guides the physician.

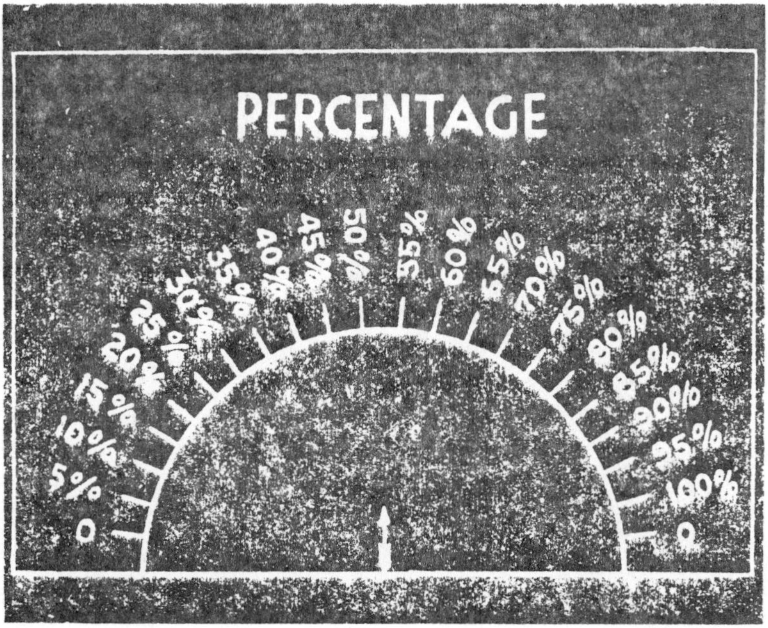

Fig. 16. Percentage Chart

Besides the Health Index and the Percentage charts there are two Number Radiation charts, and the Year, Month, Days of Week, Date and Hours Radiation charts. The number radiation charts will show, for instance, in how many days a patient will show good or bad signs, how many days a critical patient is expected to live or be out of danger. Year, month, date, day of week and hour charts will point out the time for a particular event, be it an ordinary event or death or recovery as the case may be. These charts will also indicate answers to endless questions relating to Time, and should be studied carefully. Brief notes on the charts and their usefulness are given here in this chapter.

The Health Index and the Percentage charts illustrated in Figs. 14 and 16 are of great value in magnet dowsing and in the practice of medicine. As soon as a patient comes, it would be profitable to have his photographs tested on the Health Index chart first. This test will at once show the state of his health and the probability of his getting a cure. So long as death is there in the world, there is no denying the fact that we are helpless against disease and death both of which are dependent on the higher cosmic forces. Magnet dowsing is the process by which the disposition of the higher cosmic forces can be detected with some measure of accuracy.

On the centre of the Percentage chart or the centre of the base line, the photograph of the patient or a person under test should be placed. Over this photograph a pendule should be operated in the manner previously indicated. The pendule with a pointed lower end will move by its own force to the different entries on the chart. If it shows 75%, then it should be understood that he has finished three parts of his life and only one part remains. Suppose a person at 60 shows 75% health index, then it should be understood that he has finished three parts of his life and only 20 years still remain for him to enjoy. So, there can be no question of failure in treatment. Doctor's skill will have to be exhibited where the photograph shows 95% or even 100% health index showing probably that the patient has come to the doctor for his last illness. Various other matters can be dowsed from this chart which will be indispensable for medical work. The percentage of benefit or otherwise with every dose or injection can be ascertained by a reference to this chart.

The chart illustrated in Fig. 17 is one of the many number radiation charts necessary for dowsing purposes.

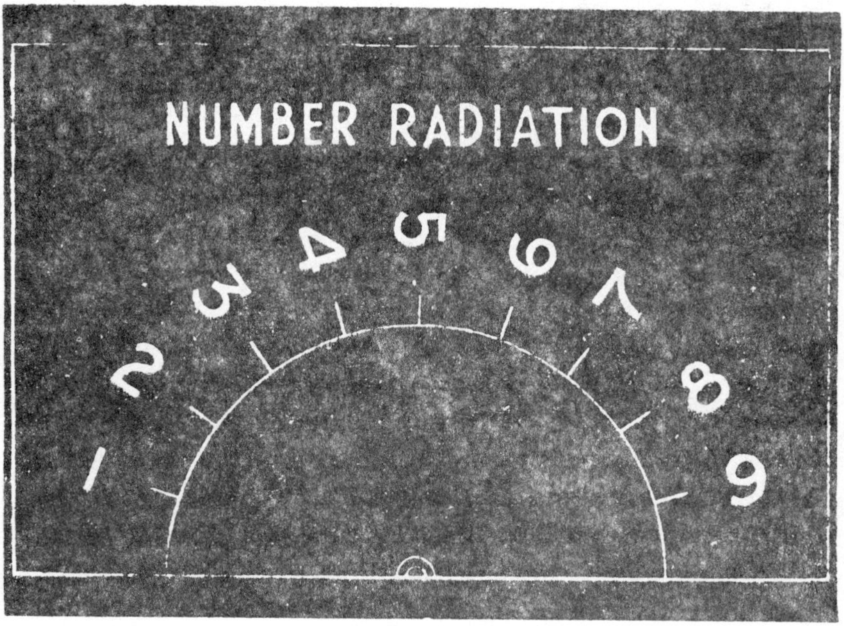

Fig. 17. Number Radiation

This chart bears the nine primary numbers from 1 to 9 and will be found useful for numerological research. This chart will show many things with regard to Time. When, for instance, a photograph of chronic patient is placed on the chart, and the pendule shows a number it means that recovery or improvement will take place when the indicated number of months have expired. In an acute case, if a question is asked as to how many days it will require to give him relief, the number pointed out by the pendule will be the number of days required for the patient to get some relief. What particular number is beneficial or otherwise to an individual will be told

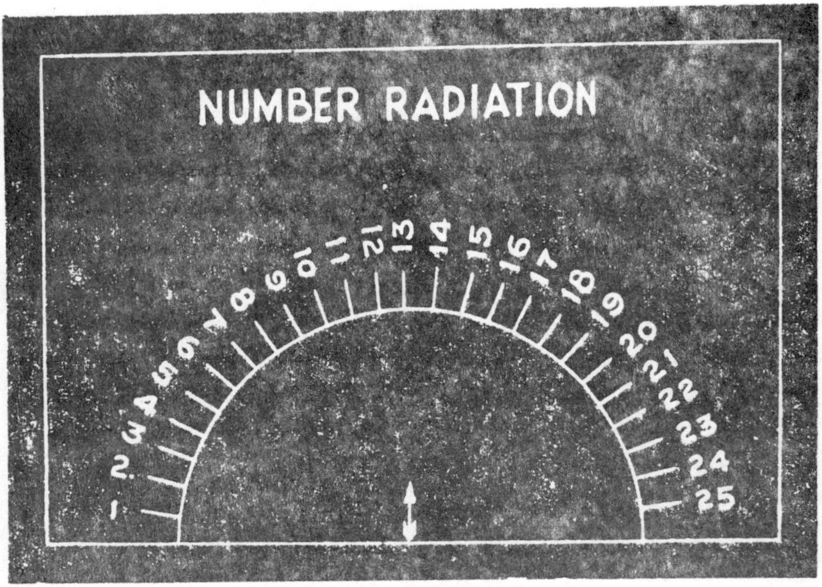

Fig. 18. Number Radiation

by this chart. What dates are beneficial or otherwise for a person will also be indicated by this chart. The number sacred to the different planets such as Sun, Moon, Mars, etc. will be indicated by this chart, and the prevailing notions about the numbers may either be confirmed or modified by the information obtained from this chart. The photograph or the name card of the person under test should be placed on the centre of the base line, and the pendule should be operated on this chart in the same manner as indicated before.

Very often it becomes necessary to find out our required information by a number which may be above 9. In the chart illustrated in Fig. 18 there are altogether 25 numbers which should be used to find out answers to numberless questions that arise in our minds. The number of months, required for a patient to recover, the number of days a patient will be on bed, the number of days a serious patient is likely to live, after how many days for instance, a near and dear relative will return to his native country or to his parents, or the number of years after which a student will return from foreign land, such and such other endless questions can be answered with the help of this number radiation chart. Even auspicious and inauspicious numbers for an individual also can be divined with the help of this chart. Such dowsing results are not meant to be communicated to others, and the replies are to be taken as approximate only and not as absolute certs. This chart should also be operated with a pendule with a pointed lower end in the manner indicated before.

The Year radiation chart illustrated in Fig. 19 is another of the several Time charts necessary for the purpose of dowsing required in every day life, as also for medical purposes. This chart shows the years in A.D. from 1950 to 1980 for thirty years. For later years new charts may be prepared and used for the purpose of magnet dowsing. The present chart may be used for all future happenings in years of the Christian era, such as birth, death, disease, accidents, appointments, dismissals, promotions, demotions, years of fire, epidemics, social upheavals, and such other endless happenings in the future. The favourable and unfavourable years for health, finance, property, children, accidents, litigations, and the rest in the case of an individual can be ascertained by a reference to this chart.

Once I had a case of cancer which I declined to take up after dowsing. The year radiation chart showed 1950 as the last year of her life, May as the last month of her life and

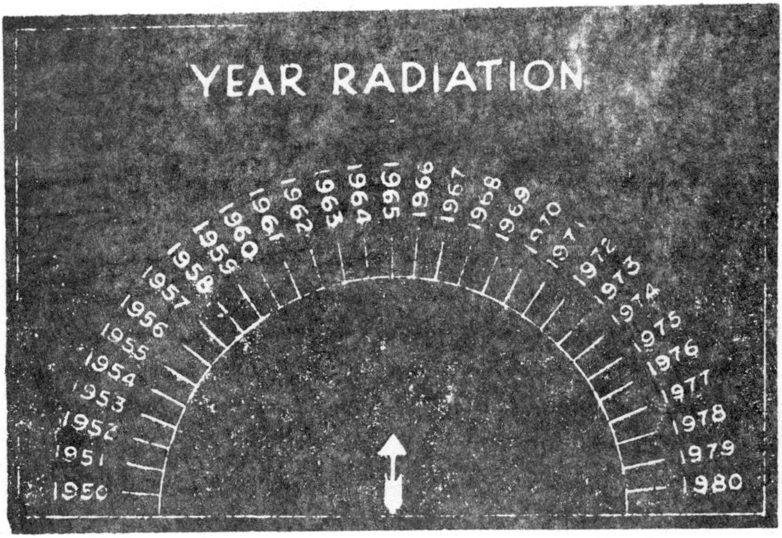

Fig. 19. Year Radiation

the last date as 21st. In April 1950 the case was referred to me, while the tests showed that the patient could not live beyond 21st May. It was difficult to believe in future in this manner also, and as the guardians of the patient insisted that I should take up the case, I told them simply that the case would be taken up after 21st May. She was apparently in the best of health with no alarming symptoms, but all on a sudden exactly on the 21st day of May, 1950 at 3 P.M. she died of heart failure and nothing could be done under the best of medical treatments. From the case cited here it should not be concluded that in every case death can be predicted like this, because the higher cosmic forces are averse to reveal this knowledge to ordinary mortals.

The method of dowsing is the same as before. On the

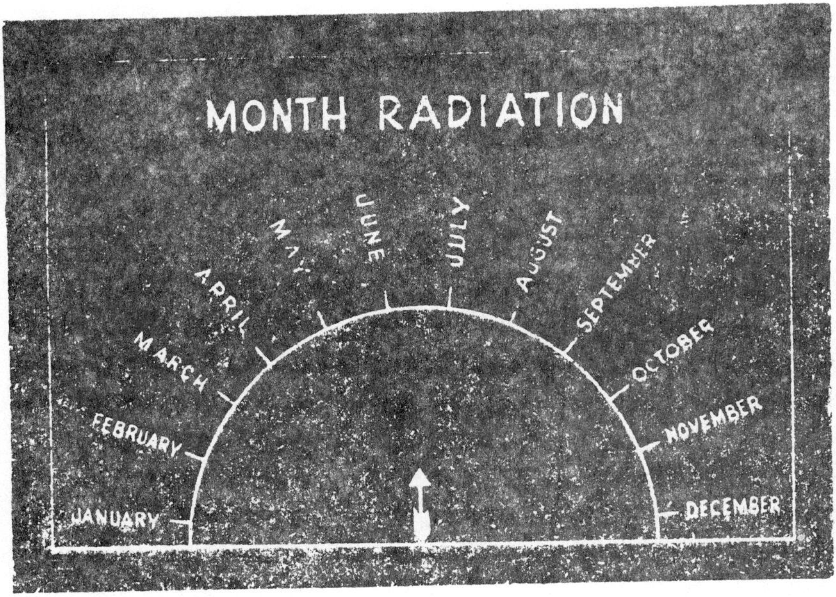

Fig. 20. Month Radiation

centre of the base line the photograph or the name card of the patient should be placed. Over this the pendule with a pointed lower end should be operated as before, and by carefully watching the pendule, the correct year should be noted.

The Month radiation chart illustrated in Fig. 20 is one of the time charts required in magnet dowsing. There are altogether twelve entries on the chart and these are the twelve months of the year. When a particular event should take place, when some one will be benefited or harmed, the month of promotions and demotions, joys and sorrows, disease and cure, honours and insults, gains and losses, passes or failures, and such other endless questions relating to month can be ascertained from this chart.

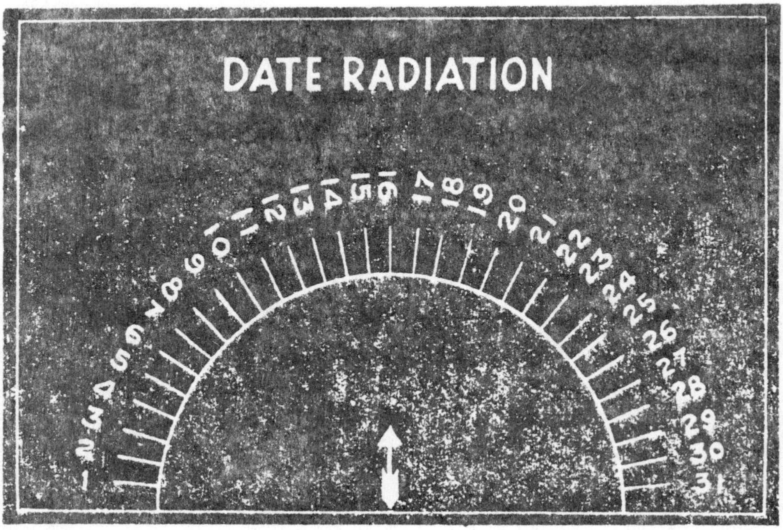

Fig. 21. Date Radiation

The method of dowsing is the same as before. The photograph or the name card of a person under test should be placed on the chart on the centre of the base line. Over it a pendule should be operated in the manner aforesaid and by carefully watching the movement of the pendule the required month should be fixed. Repeated tests are required for all the Time charts dealt with in this chapter.

In life, it often becomes necessary to find out the date on which a particular event should take place. The information is required both for ordinary and medical purposes. This time chart, illustrated in Fig. 21 will prove useful whenever it becomes necessary to divine the date on which an event would take place. On what date, for instance, a patient should feel better, the date on which there will be remission of fever,

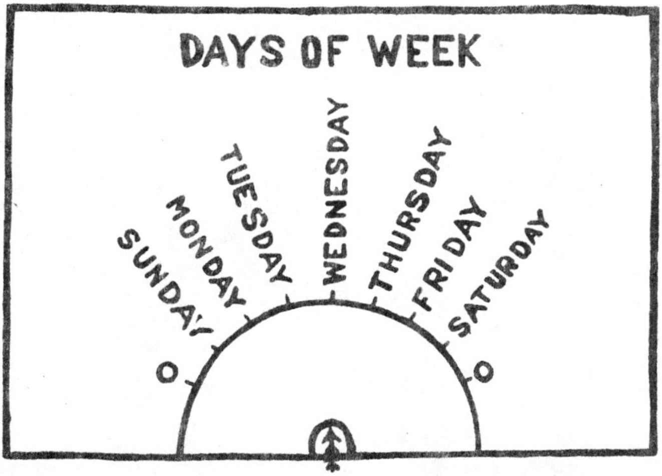

Fig. 22. Days of Week

upto what date a patient will have to suffer, on what date first diet should be given and such other endless questions can be answered by this Date radiation chart. The pendule should be operated in the same way as before and it will be better if a pendule with a pointed lower end is used.

Next in the order comes the Days of the week chart (Fig. 22). On this chart the auspicious or the inauspicious days of the week can be found out for a particular person. It may be of some interest to readers to know that each day is under the influence of a particular planet. Sunday is ruled by the Sun, Monday by Moon, Tuesday by Mars, Wednesday by Mercury, Thursday by Jupiter, Friday by Venus and Saturday by Saturn. For each individual some days are beneficial, as the planets ruling those particular days of the week are beneficial to him. It is always helpful to know on what day a particular work should be undertaken etc.

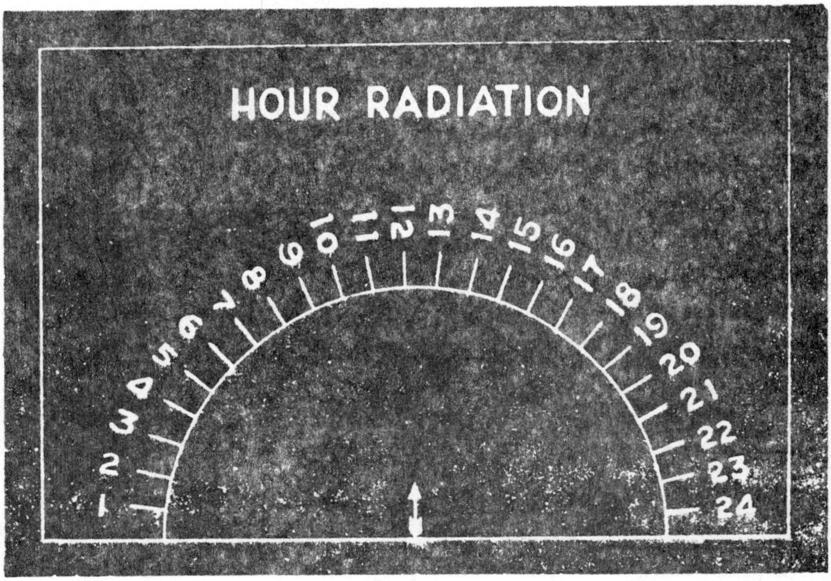

Fig. 23. Hour Radiation

The chart illustrated in Fig. 23 is an ordinary time chart for common and medical use. When, for instance, it becomes necessary to ascertain in how many hours a patient will improve or will become worse, at what hour a particular event should take place, or when it is to be ascertained what hours are good or bad for an individual or a patient, or upto what hours a medicine or injection will have its action, or at what hour medicine should be repeated, and such other endless questions can be answered by this chart.

Each hour of the day and night is again dominated by a particular planet. The first hour after sunrise is ruled by the planet that governs the day. The order of motion beginning with the planet of the longest period is as follows: Saturn,

Jupiter, Mars, Sun, Venus, Mercury and Moon. For example, Monday is ruled by Moon in the first hour, 8th hour, 15th hour and 22nd hour. Similarly, according to the above sequence, the planet ruling a particular hour of a particular day can be found out From the chart the good and bad hours of every day for an individual can be found out with ease.

Here also the entries are too many and for accurate work a small pendule with a pointed lower end will be necessary. The photograph, medicine phial or the name card of a person may be placed on the centre of the base line, and the pendule should be operated in the manner indicated earlier. By observing the movement of the pendule the required hours may be fixed. To obtain mastery over these Time charts much practice will be necessary. The operator may be cautioned that from these tests too much should not be expected, and all conclusions regarding the future may or may not come true. Under no circumstances the results are to be made public or communicated to others.

CHAPTER VII

Magnet Study of Disease

Magnet dowsing is of some assistance in the diagnosis of disease. It will also be profitable to know all about the disease and the cosmic factors involved in each and every disease. Without this valuable guidance the true value of the great destructive power of disease will be rather difficult to comprehend. The charts dealt with in this chapter relate to different diseases and have been described in the sequel with brief notes which are likely to be of value to all sincere inquirers.

First of all, it must be known to what particular element a disease belongs. There are altogether five cosmic elements which constitute the cosmos along with its mobile and immobile contents. The five elements are named in philosophical works as Earth, Water, Fire, Air, and Ether. Diseases belong to one or another of the five elements just mentioned. If the disease belongs to the earth element, it calls for a medicine which also belongs to the earth principle. If the disease belongs to the element of water, it calls for a medicine belonging to the identical element. Diseases, where the ether element is out of order, are generally the most difficult, and the diseases where the element of air is deranged, are the most painful ones requiring great skill in treatment. The relation between the disease and its originating elements has not been given the attention it deserves. Magnet dowsing can help in establishing this connection.

The five elements have contributed one or another of the five Vayu forces, well known as Prana, Apana, Samana, Udana and Vyana. These Vayu forces are contributed by the elements of Earth, Water, Fire, Air and Ether respectively. This Vayu chart should be consulted in order to verify the element, and select a medicine which covers both the element and the Vayu force equally.

Another chart included in this chapter will help, with a little care and attention, in finding out the seat of the disease for every individual. The information derived from this chart will also prove of assistance to the prescriber to some extent, but the element and the Vayu force are to be considered more important than the seat of the disease for the purpose of treatment.

The Condition radiation chart is another important instrument which will prove of value to the prescriber. If the disease is moderately difficult, it will show ill, very ill or bed-ridden, but if the disease is a killing one and very dangerous the chart will show serious, critical, hopeless or dying condition, and in all such cases the disease should be considered dangerous or very nearly fatal, and great care is required to be taken of such patients.

Ancient philosophers have declared that the macrocosm or the universe is composed of the five elements, earth, water, fire, air, and ether. In the Upanishada of Ancient India the origin and order of the elements are given as follows. First of all, from the Supreme Power of the universe originated the element of ether which was followed by the element of air ; from air arose the element of fire which was soon followed by the element of water, and from water came the element of earth. These elements have their subtle counterparts of qualities and they are known as sound, touch, sight, taste and smell successively. From these subtle counterparts of the five elements originated the five faculties in human and other sentient beings, namely, hearing, feeling, seeing, tasting and smelling through the respective sense-organs such as ears, skin, eyes, tongue and nose.

The five elements pervade the whole universe and they contribute their own forces which are manifest everywhere in creation. Thus the element of earth contributes the prana (or the Vital force), the element of water the Apana Vayu (or the downward force), the element of fire the Samana Vayu (or even force), the element of air the Udana Vayu (or the upward force) and the element of ether contributes the Vyana Vayu (or the universal force). For a detailed description and the functions of these five kinds of Vayus or forces Ayurvedic works may be consulted. A reference may also be made to the present author's 'The Science of Tridosha'' published by Messrs Gotham Book Mart, New York. As these facts are well-known, it is not desirable to deal with them extensively here.

When a person is suffering from a disease, it is necessary to find out to what particular element that disease is related. Magnet dowsing will make it easy to ascertain this all-important relationship. Even a disease belongs to one or another of the five elements and it is desirable to know to what element a particular disease is related. For instance, when there is emaciation, it is to be surmised that the disease belongs to the element of earth. The earth element in the human system is represented by its heavy contents such as skin, bone, flesh, fat, viscera, and the rest. The force of the earth element is the Prana Vayu by which food taken through the mouth goes into

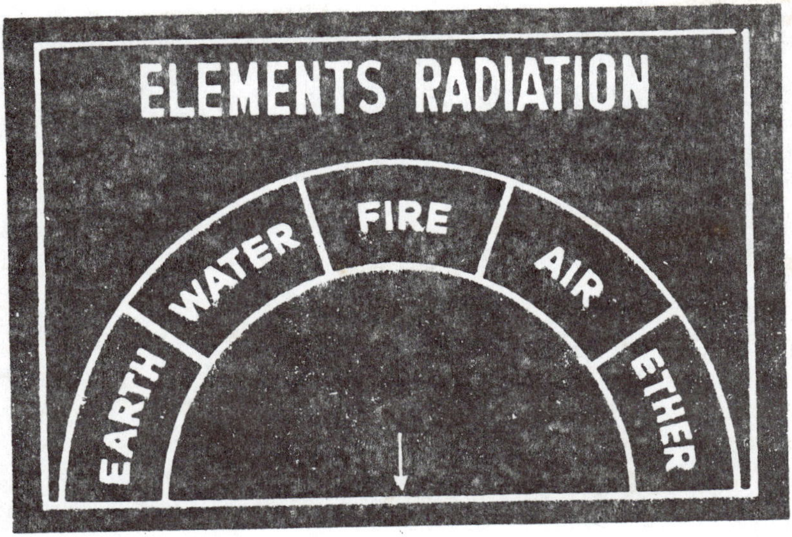

Fig. 24. Elements Chart

the stomach, and with the help of which we breathe air and take it to the lungs and throw it out in the reverse manner. When emaciation takes place, the patient loses flesh, and thus the element of earth in the body becomes deficient. This is usually the work of heat. Similarly, tuberculosis, polio and suchlike diseases also affect the element of earth. The relation between the disease and original cosmic element can be ascertained by a reference to the Elements radiation chart illustrated in Fig. 24.

The element of water is involved in various diseases of thin and thick lymphs and secretions and blood circulation. Even insanity, epilepsy and such diseases come under the element of water. Fear is one of the principal symptoms when this element is deranged in the human body. Bad liver and spleen always indicate impairment of the water system. Usually photograph of such patients will show the attraction of the pendulum towards the water element in the chart, and the disease will also indicate that the water element is out of order, requiring medicines which have affinity with the water element.

The element of fire, when deranged, will give such diseases as rickets, piles, anaemia, lukaemia and various other diseases involving the fire of digestion and the power in the marrows for the formation of new blood. Weak eyesight without injury to the material part of the eyes comes under the fire element. Diseases of the fire element will require fire medicines and vibrations of Ruby and Coral both of which are hot.

The element of air when deranged, gives rise to inflammation to the nerves and various other nerve diseases. Neuritis, neuralgia, neuralgic pains anywhere in the body will be readily relieved by the application of air remedies for the pacification of the Udana Vayu (or the upward Air). The element of air resides in the violet cosmic colour and in the gem, Sapphire.

The element of ether which resides in every part of the body and in the hollow passages with which the body abounds, when deranged, gives rise to various diseases of the Ether group. Heart troubles, palpitations, paralysis, anasarca, oedema, swellings, whooping cough, and diseases caused by small bacteria are the diseases relating to the element of ether. At the time of death, the element of ether and the Vyana Vayu are the last to be deranged. The planet Jupiter is the lord of the Vyana Vayu and the element of ether and is the giver of life. The gem which has Vyana Vayu in it is the Moonstone (white or yellow Pusparaga), or Topaz or crystal.

In Teletherapy these gems are rotated on electric motors in order that these gems release a great volume of Blue cosmic colour which gives strength to the Vyana Vayu. Goitre, elephantiasis and suchlike diseases are the result of a deranged Vyana Vayu, and certainly Blue is the colour which can restore the balance of the Vyana Vayu by strengthening it.

The elements radiation chart thus is an important chart and may be consulted frequently by all those who are interested in the origination of disease and cure. Dowsing on this chart has to be done in the way shown earlier. First of all, on the centre of the chart that is to say, the centre of the base line a photograph of the patient should be placed. A pendule should be operated over it in the manner previously indicated, and by closely watching the movement of the pendule the element involved should be determined.

Similarly, a medicine phial may be kept on the centre of the base line over which the pendule should be held. By operating the pendule it can be ascertained as to the element to which the medicine is affiliated. This is the process by which the element imbedded in every medicine, known and unknown can be found, and this information will certainly prove of value to every one connected with disease and its eradication.

The procedure is somewhat different when the elements of diseases have to be ascertained. For this purpose when a disease is to be tested in this manner, it will be necessary to type out a card bearing the name of the disease which should be placed on the centre of the base line of the chart. The pendule should be operated over this card in order to find out the particular element to which the disease is related. Repeated tests should be made before a decision can be taken, because these constitute the most delicate of all dowsing tests.

In England and U.S.A. Witnesses[1] are available which may also be used instead of the name card of the disease (MacBeth of England is the agent for Witnesses made in France by Turenne).

Below is shown in a tabular form the relationship of the elements with various other cosmic and other forces.

Element	*Quality*	*Sense-organ*	*Planet*	*Gems*	*Colour*	*Vayu*
Earth	Smell	Nose	Mercury	Emerald	Green	Prana
Water	Taste	Tongue	Moon	Pearl	Orange	Apana
Fire	Sight	Eyes	Sun	Ruby	Red	Samana
Air	Touch	Skin	Saturn	Sapphire	Violet	Udana
Ether	Sound	Ears	Jupiter	Topaz	Blue	Vyana

N.B.—Mars is subsidiary Fire. Venus is subsidiary water.

It has been pointed out already that the five Vayus are the five forces contributed by the five elements, earth, water, fire, air and ether. Those who deal with diseases and their cures are required to ascertain very accurately as to whether a disease belongs to one or another of the five primordial elements. For instance, if the disease be typhoid, tuberculosis, small-pox or measles it is to be understood that here the element of earth is deranged along with its Vayu force which is known as the Prana Vayu (or the Vital air). If it is asthma, it is the element of water that is involved with its Apana Vayu or the Downward Vayu. If it is piles or bleeding from orifices it is the element of fire that is deranged along with its force of Vayu which is called the Samana Vayu (or the Even air). If it is neuritis the element of air is involved along with its Vayu force which is known as the

1. Witness is a word used to mean samples of organs etc.

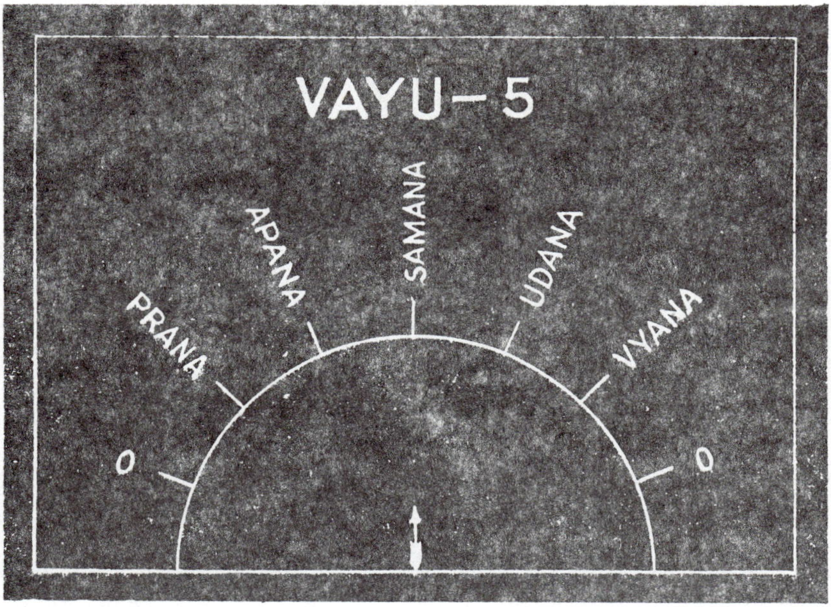

Fig. 25. Vayu--5 Chart

Udana Vayu or the Upward air. If on the other hand the disease is tumour, swellings, oedema or anasarca the element involved is the Ether along with its Vayu force which is known as Vyana Vayu (or the universal air).

For ascertaining the deranged Vayu in a given case the chart illustrated in Fig. 25 may be consulted with advantage. Except by magnet dowsing it is not possible to ascertain easily the element or its Vayu force involved in a given case. The importance of magnet dowsing under similar circumstances cannot thus be over-rated.

Once we are able to ascertain correctly the offending Vayu, it is not difficult to arrive at the correct cosmic colour hunger. Prana being the Vayu of the earth element shows Green hunger ; Apana being the water element its colour hunger is either Orange or Indigo ; Samana being the fire element its colour hunger is either Red or Yellow ; Udana being the element of air its colour hunger is Violet, while the Vyana Vayu being the outcome of the element of ether its colour hunger is Blue. When two colours are indicated it should further be clarified by a reference to the colour chart given elsewhere. Tests on the colour chart will enable one to fix the single colour.

The method of testing is the same as before. On the centre of the base line the photograph of the person under test should be placed. Over the photograph the pendule should be operated in the manner previously indicated, and by observing the movement of the pendule the correct Vayu should be ascertained. The same method applies to medicines and diseases.

Diseases usually attack one or another of the seven systems of the human body, and in order to find out the particular system involved the system radiation chart illustrated in Fig. 26 will be found to be of more than usual interest. According to Ayurveda there are altogether seven clear cut systems in the human body and they are known successively as Chyle, Blood, Flesh, Fat, Bone, Marrow and Sperm (Semen). It is very essential for a physician to know what particular system is involved in a disease, if he wishes to tackle the disease intelligently.

The systems radiation chart with seven entries is designed to help such of the physicians who are eager to check their own knowledge with the help of magnet dowsing, and find out the correct system in which the particular disease resides. In

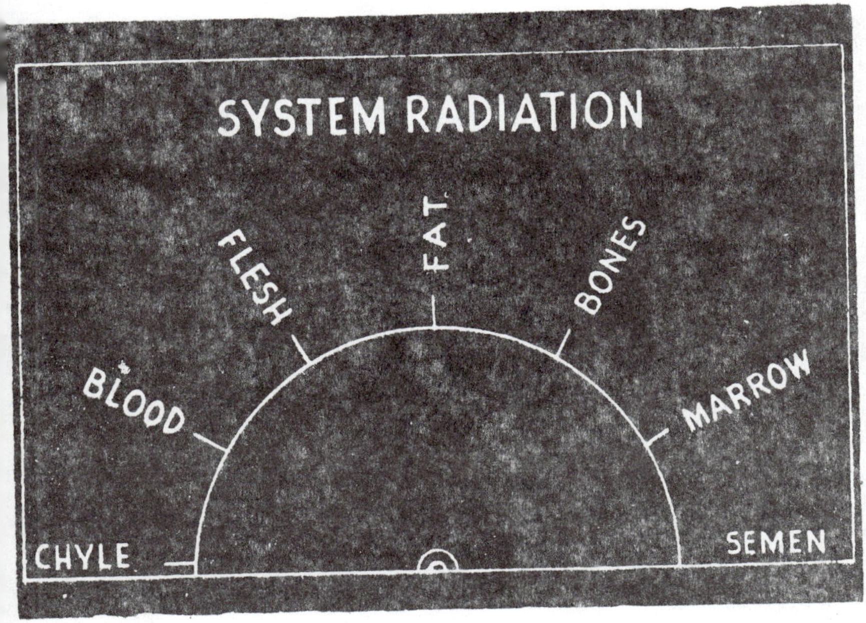

Fig. 26. Systems Chart

the same way with the help of the medicine phials and disease cards their related systems can also be ascertained.

The seven systems can be connected with their relative cosmic colours. The Chyle and Blood systems are under the influence of the Orange cosmic colour, the Flesh system is under Green, the Fat system is under the influence of Blue, the Bone system is under the Red cosmic colour, the Marrow system is under the Yellow cosmic colour, while the Sperm system is under Indigo. The nerves are here not recognised as a separate system but they are under the influence of the Violet cosmic colour.

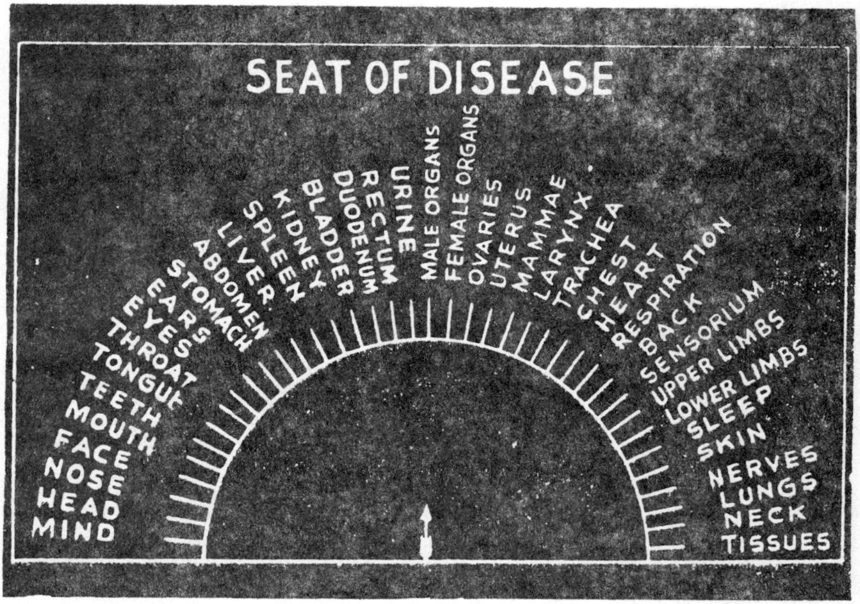

Fig. 27. Seat of Disease

For the purpose of dowsing the photograph of the patient should be placed on the centre of the base line of the chart, and a pendule should be operated in the manner indicated earlier, and by observing the movement of the pendule the correct system may be determined. For finding out the systems related to medicine, phials or tubes of medicine should be used instead of the photograph. With the help of name cards of disease, their related systems may also be divined.

When it becomes necessary to find out the seat of the disease in any given case the chart illustrated in Fig. 27 will be found to be of interest, and it should be consulted whenever there is any doubt about the seat of the disease in the body.

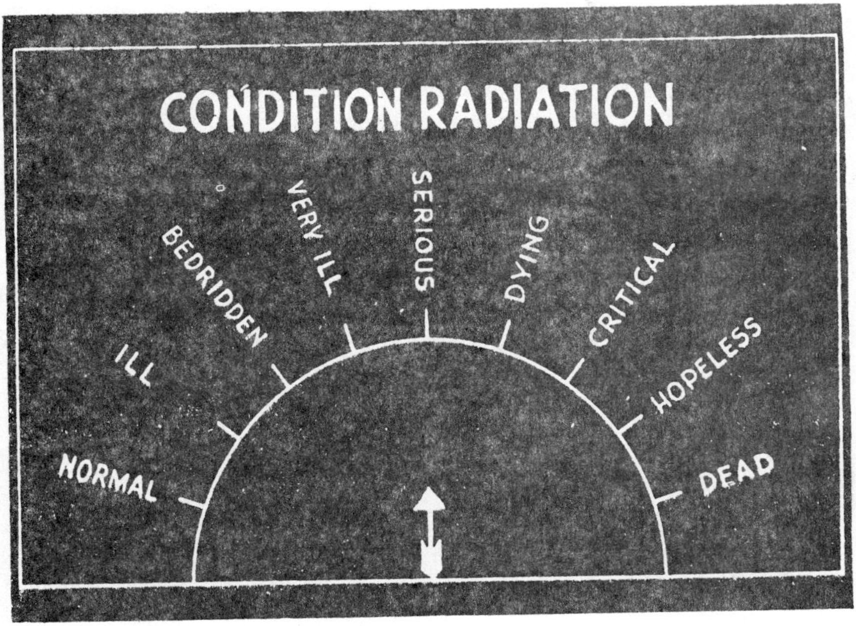

Fig. 28. Condition Chart

This chart has many entries according to various important human organs, and great care is necessary to conduct tests on this chart. First of all, it is necessary that instead of the magnet, a silver pendule with a pointed lower end should be operated with the right hand, and the chart should be magnetised with a second magnet. This magnet to magnetise the chart is not essential when dowsing is done on the Electro-magnet Box.

The method of dowsing is otherwise the same as before. On the centre of the base line the photograph of the patient should be placed, and the pendule should be operated in the manner indicated previously, and by observing the movement of the pendule the right organ should be ascertained. It is a difficult test and much practice is required before reasonable mastery over the chart may be secured.

Often when a patient has to be treated from a distance it becomes necessary to know the condition of the patient. This can be done by magnet dowsing. With the help of the Condition radiation chart illustrated in Fig. 28 it is possible to ascertain, for instance, whether the patient at the moment of dowsing is well, happy in a diseaseless state or whether he is ill, or very ill, or in a serious, critical or dying state. This chart has also to be used with great care.

The method of dowsing is simple as before. On the centre of the base line a photograph of the patient should be placed. Over this photograph the pendule should be operated in the manner indicated previously and by observing the movement of the pendule the correct condition of the patient at the moment can be determined. Information obtained by this test is not meant for communication to others.

It is also essential to know to which plane the disease belongs. This can be found out from the Plane Chart as shown in Fig. 29. There are three entries namely Spiritual, Psychic and Physical. When the disease belongs to the Spiritual Plane, it means that no definite cause can be found for the particular disease and it will be a difficult task to cure it. Treatment of a subtle nature is required in such cases. A Psychic plane disease means that the person's thoughts are responsible for the particular disease. For example, in a person suffering from diabetes, if the psychic plane is indicated, it may be taken for granted that the disease has originated from worry and anxiety, which is one of the

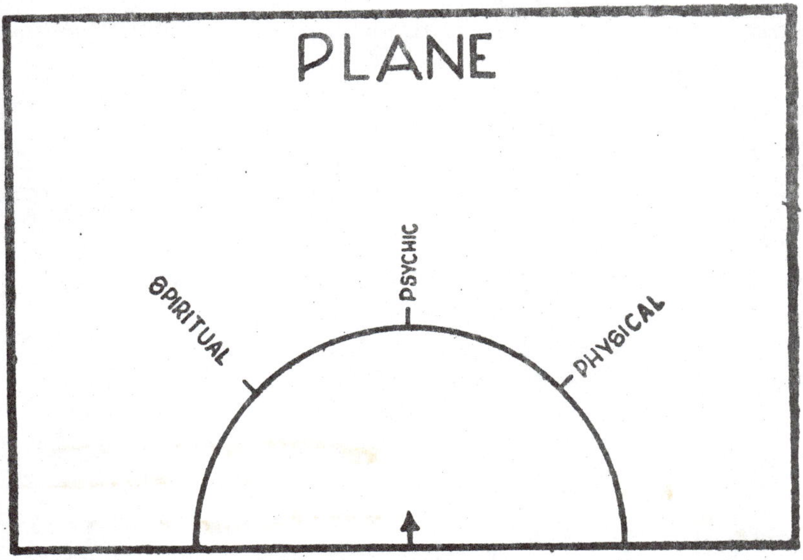

Fig. 29. Plane Chart

causes of diabetes. The Physical plane indicates an outside agency, such as, injury, infection, etc. The above information is very helpful, as it helps in the selection of the colours for the person. The colours belonging to the Spiritual plane are Red, Blue and Violet. The Psychic plane colours are Orange, Green and Indigo. Lastly, Yellow, Ultra-violet and Infra-Red colours belong to the Physical plane.

The next three charts are Brain Condition (Fig. 30), Brain Heat Percentage (Fig. 31) and Brain Cold Percentage (Fig. 32). In any disease in a person where the brain or the Nervous System is involved, these charts should be consulted to find out the condition of the brain. The first chart on which dowsing should be done is the Brain Condition Chart. After finding out whether there is Heat or Cold in brain, the next chart of either Brain Heat or Brain Cold Percentage should be consulted. Here also the principles followed when investigating with the Health Index or the Percentage Chart are applicable. If the reading is 70% or below, then the condition is not grave, but as the reading goes up, the condition becomes serious. A reading above 95% may be considered very grave and a cautious prognosis should be given regarding the case. In many serious cases the brain is involved and these charts help to find out the immediate condition. Hence, these should always be consulted in all serious cases.

The various charts used in this book may be used for confirmation of other results of dowsing. The Radiation Required chart (Fig. 33) may be used to find out which colours are indicated in a given case. The Hot colours are Red, Yellow, Blue and Infra-Red, while the Cold colours are Orange, Green, Indigo, Violet and Ultra-violet. In dowsing with the Brain charts if the finding is Heat in brain, it automatically means that Cold colours or medicines are required. But if in dowsing on Radiation Required chart the findings are Hot Radiation, then it means that the dowsing results are wrong either with the first testing or with the second testing. These dowsing tests should be repeated several times before coming to a conclusion.

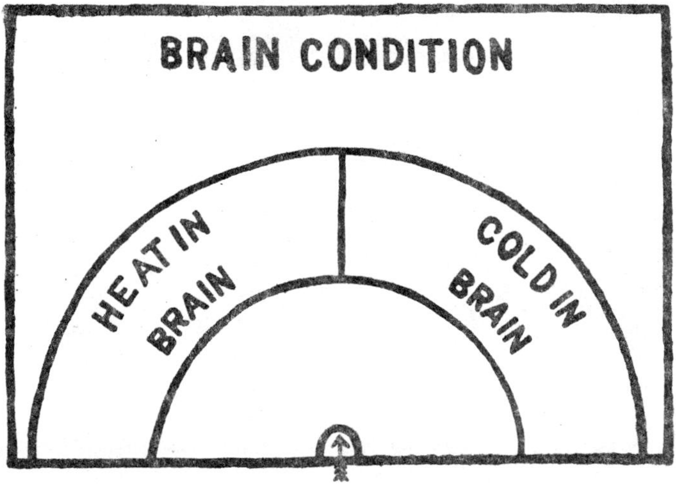

Fig. 30. Brain Condition

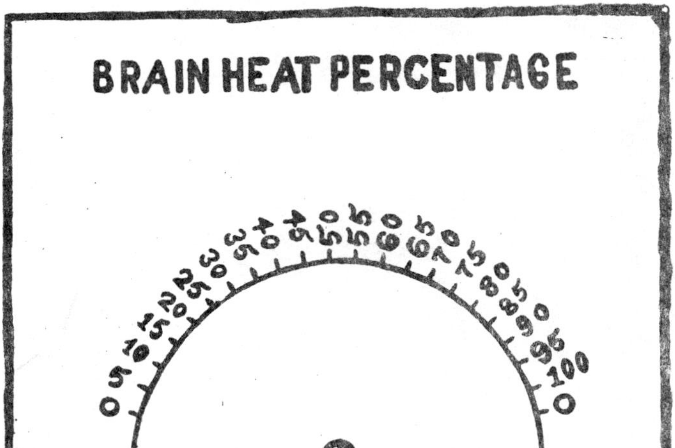

Fig. 31. Brain Heat Percentage Chart

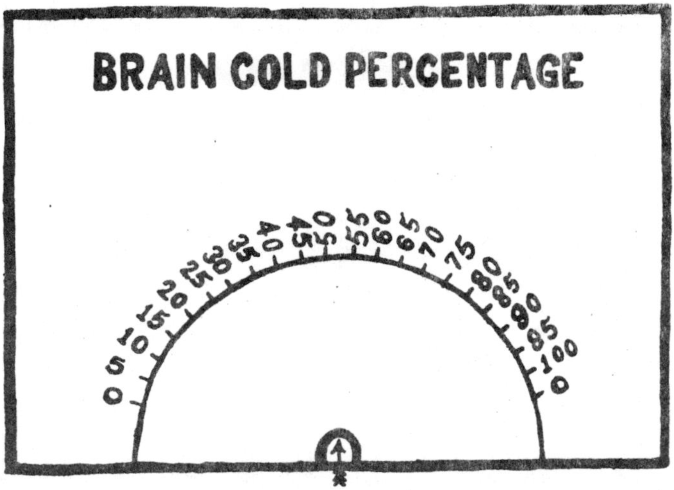

Fig. 32. Brain Cold Percentage Chart

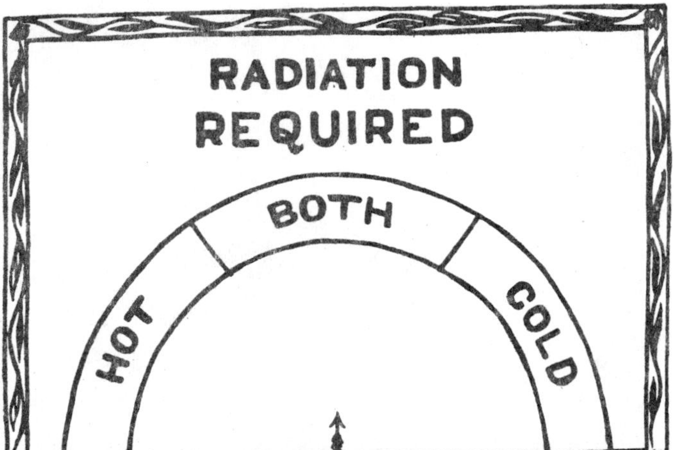

Fig. 33. Radiation Required Chart

CHAPTER VIII

Magnet Study of Medicine

Just as an individual can be studied and analysed by magnet dowsing, even so each and every medicine, known as well as unknown, can be tested under different heads in order to ascertain its quality and power, and to determine its usefulness to individuals. The very first thing to be known in a medicine is its taste, and unless this taste is definitely fixed for a medicine it cannot be correctly employed in the alleviation of human suffering. The taste radiation chart is designed for this purpose, and the magnet by moving over the medicine kept on the chart will show the particular taste to which the medicine belongs. Both for old and new medicines this taste can be ascertained, but without knowing the taste it is not possible to determine whether the medicine has effect on Vayu (air), Pitta (fire) or Kapha (water). The taste of Homoeopathic medicines has not yet been ascertained, and it is of utmost importance to investigate this question before the true powers of the medicines can be gauged. This equally applies to all medical systems which use untested and scientifically undetermined new medicines.

Whether a medicine should be used in treatment or not is to be ascertained by the next chart which tests the stage to which the medicine belongs. This chart will indicate whether a medicine belongs to the Child stage or Youth stage or to the Dying stage. Those medicines that belong to the Youth stage are the best. Boy stage is the next best, while others are just

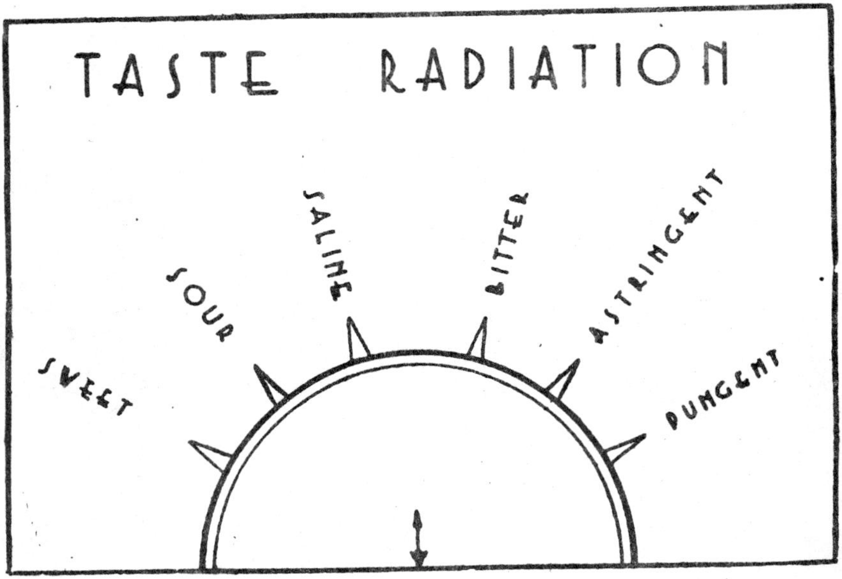

Fig. 34. Taste Chart

medicines. Medicines belonging to the old and dying stages are definitely bad and, as far as possible, should be avoided. This stage of medicines can be tested in relation to individuals in order to find out the relative suitability.

There are other charts of equal value and they will be found useful in divining Biochemic medicines for individuals and for diseases, in ascertaining the potencies of homoeopathic drugs for different purposes and for different individuals, and in determining the interval of time between doses for different medicines and individuals.

The Science of Ayurveda recognises six different tastes for all conceivable medicines and items of food. These six tastes build the tangible body of the different sentient beings.

Excess or shortage of tastes is said to be one of the causes of bad health. These six tastes are well known as : Sweet, Sour, Saline, Punget, Bitter, and Astringent. These are the six entries in the chart under discussion and illustrated in Fig. 34.

With the help of this chart it will be possible to find out the taste of each and every medicine used by men, and if the tastes are known their qualities also become manifest. According to Ayurveda every taste has its definite action of amelioration and aggravation of the three cosmic elements such as Vayu, Pitta and Kapha.[1] It is given in Ayurvedic works that the three tastes, Sweet, Sour and Saline pacify Vayu (air). Sweet, Bitter and Astringent pacify Pitta (fire), while Pungent, Bitter and Astringent pacify Kapha (water). Without ascertaining the different tastes inherent in every possible medicine of this, that and other schools it is not at all possible to determine their true qualities. Sometimes I wonder how conscientious doctors can use their medicines without knowing the tastes inherent in them !

The homoeopathic medicines are potentized medicines, and they lose much of their substance and their tangibility in the process of potentization. In spite of this it should be remembered that every one of these medicines has a taste and with the power of this taste in potentized form it is able to cure complicated diseases. But the tastes of these valuable medicines have not been investigated nor ascertained. This is a stupendous task, but it should not be neglected any longer in view of the supreme importance of this investigation.

---

1. All about the Tridosha or the working of the Three Cosmic Elements in the human body will be found described in brief in my work, entitled, THE SCIENCE OF TRIDOSHA, published by Messrs Gotham Book Mart of New York (1856, 2nd. edition).

Magnet dowsing will help investigators in ascertaining the true taste of all medicines used in Homoeopathy. It may be remarked here that in Ayurveda and Unani systems there is not a single medicine whose properties are not known or whose taste had not been determined long long ago.

This taste radiation chart will prove useful to all those who wish to investigate the taste of medicines as well as foods in order to ascertain their quality and power through magnet dowsing. In order to ascertain the taste of a medicine, say, Capsicum which is pungent in quality, the homoeopathic medicine phial or a long pepper may be placed on the centre of the base line of the chart. The pendule should then be operated over the phial or pepper, and by closely watching the movement of the pendule it will be found that the pendule is attracted towards the entry on the chart which is marked 'Pungent'. Likewise, other medicines or foods will show their respective tastes. These tests are required to be made with great care. The tests should be repeated several times before a decision is taken about the taste of medicines and foods. If by chance there are many investigators in the field practising magnet dowsing their results should be compared and thus a balance should be struck in the matter of taste of individual drugs and foods.

The Stages Chart illustrated in Fig. 35 will appear to be of more than usual interest, and can be used in various ways. Whether a disease is in an early or mature stage, whether it is youth or dying stage, that is to say, whether the disease will lead to death or not, and such other questions can be dowsed with the help of this chart. A medicine can also be tested on this chart for its efficiency and quality. If it is in the stage of a child, it is not efficacious, if it is in the

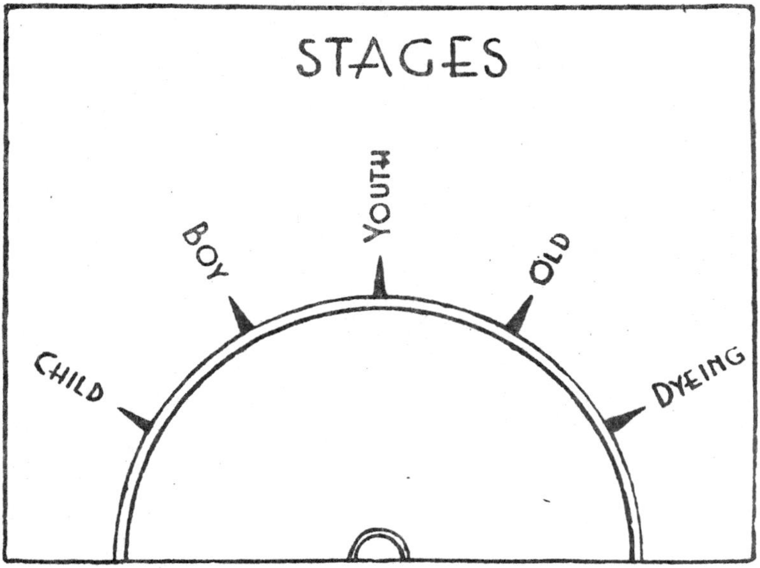

Fig. 35. Stages Chart

Boy stage, it is moderately efficacious, but if it is in the Youth stage it is highly efficacious. In the old and dying stages the medicine is likely to be injurious. When a new medicine is prepared, it should be tested on this chart, and it should be accepted only if it is in the Youth stage. Seven Biochemic medicines develop Youth in the 6X potency, and therefore, Biochemic mixture should be prepared in this potency only. Other new drugs can be tested in the same way before putting them to use.

The method of testing is the same as before. On the centre of the base line of the chart should be placed the photograph of the patient, or the name card of disease or the new medicine. Over one or another of these the pendule should be operated in the manner indicated before, and by observing the movement of the pendule the stage of the

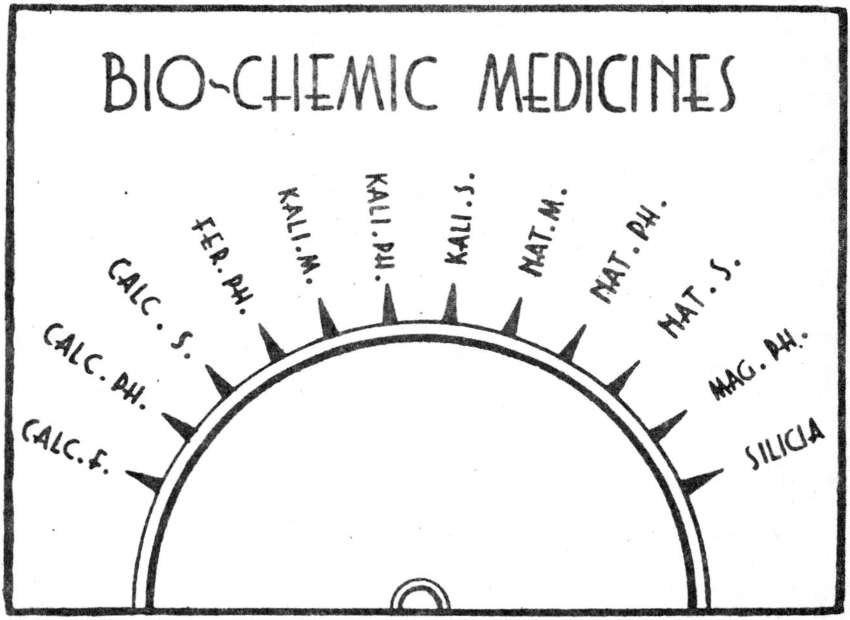

Fig. 36. Bio-Chemic Medicines Chart

medicine should be ascertained. The same test should be repeated several times before a decision is taken for all practical purposes.

Those who are interested in finding out the true Bio-chemic medicine for individuals will find the chart of Bio-chemic medicines illustrated in Fig. 36 of some practical value. The chart contains the names of the 12 Tissue remedies arranged in a semi-circle for dowsing purpose. Those who handle Biochemic medicines for a large number of patients know very well how difficult it is to select the correct remedy, even though their number is not more than twelve. Those who do not mind devoting a little of their time in dowsing the correct remedy for every individual patient are advised to use this chart for dowsing the correct medicine.

The method of selection is the same as before. On the centre of the base line of the chart the photograph or the name card of the patient should be placed. Over this card the pendule should be operated as before. By observing the movement of the pendule the name of the correct Biochemic medicine should be ascertained. For this kind of dowsing a pendule with a pointed lower end should be used. Once the medicine is fixed for an individual from this chart, the phial of the same medicine should again be tested against the photograph to see whether the phial attracts the pendule or not. If the pendule is attracted to the medicine phial placed on the Magnet Box, it should be prescribed without fear, but if the pendule runs parallel to the phial, the medicine is not indicated for the individual against whose photograph the phial is tested. As usual, tests should be repeated several times before a decision is taken. This method of testng for Biochemic as well as other medicines is strongly recommended to the profession, because by this the author hopes, that many wrong prescriptions may be avoided to the great relief of the suffering public.

The next few charts will be useful to Homoeopathic physicians who would like to investigate on these lines of Radi-esthesia. The first chart in the series is the Miasm chart, Fig. 37 which is a very important chart for the Homoeopathic physician. He should know which Miasm or Miasms are affected in a given case and this chart will help him to find this out. Dowsing is to be done on this chart in the ordinary way as explained earlier.

Homoeopathic Materia Medica contains more than 2000

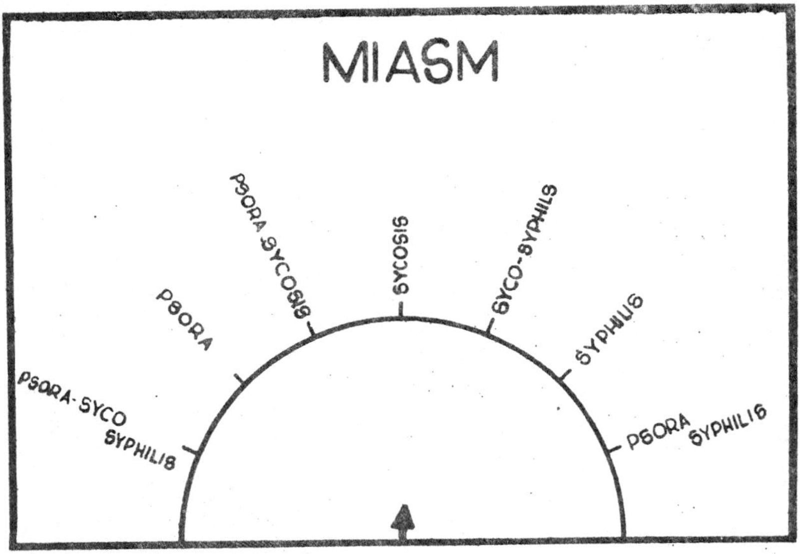

Fig. 37. Miasm Chart

remedies and it is a herculean task to find out the correct remedy from this larger number. Another difficulty is that all the medicines are not mentioned in the repertory under the heading of a particular disease and it is quite possible that some unmentioned remedy may be required in a given case. For example, Ammon Carb. is not mentioned as a cancer remedy, but the author has cured a number of cases with this remedy which he found from dowsing. As said earlier an Alphabet Chart (Fig. 38) is given which will help busy practitioners to find out under which letter of the Alphabets the particular medicine for a given case will be found. After finding the letter all medicines beginning with that letter should be tested and the correct remedy should be found out. Suppose, the letter indicated is 'M', then all medicines beginning with the letter 'M' from Macrotin to Myrtus

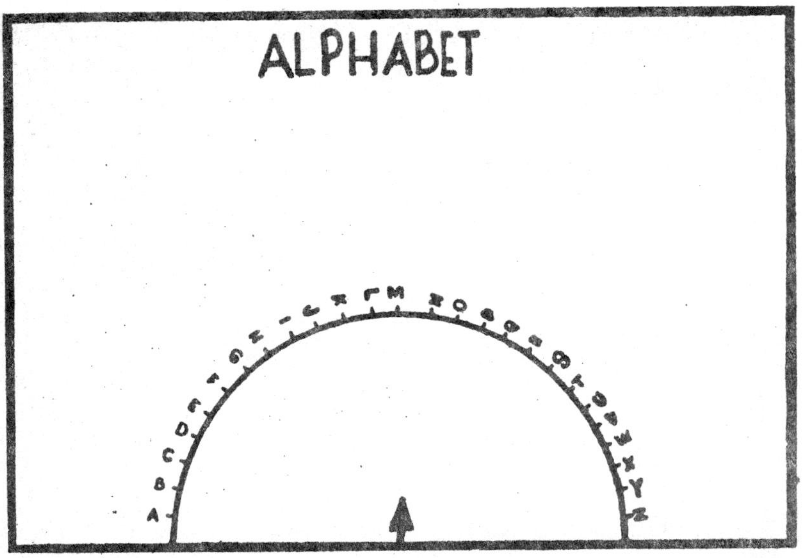

Fig. 38. Alphabet Chart

Communis should be tested for affinity with the given case and that which has the attraction is the correct remedy for the person.

Next comes the problem of selecting the potency of the medicine for the person. This also is not an easy job. Every physician has his own fancy regarding the potency. In order to be sure of the potency of the medicine, the Potency Radiation Chart—Fig. 39—should be consulted. Once the potency is fixed by dowsing, the Interval Radiation Chart as shown in Fig. 40 should be used to find out after what interval the remedy should be repeated. When all the data are found it should be double checked on the Percentage Chart and if it shows 100% then the tests are correct and the medicine thus found, may be given to the patient with confidence.

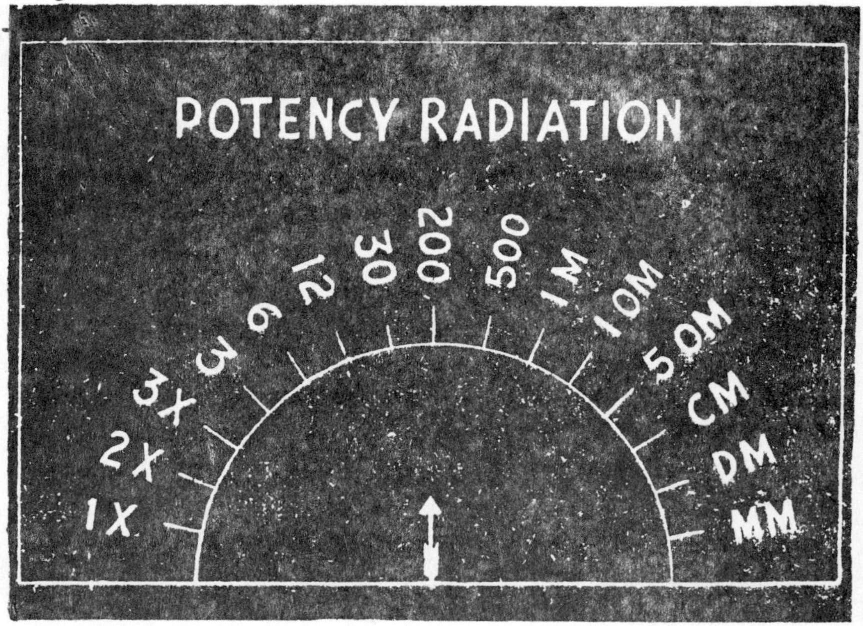

Fig. 39. Potency Radiation Chart

I will give below an example of how to do it correctly. The photograph of Mr. A is placed on the Magnet Box (either Electro-Magnet Box or the Permanent Magnet Box) and on the Alphabet Chart dowsing is done to find the alphabet from which the indicated medicine will be found. Suppose, the alphabet is 'P'. The next step is to take all the medicines under the alphabet 'P' one by one, and test their affinity with the photograph of Mr. A. As all the medicines may not be available, their name-cards should be placed instead of the medicine bottle. After testing all the medicines under 'P', suppose the indicated medicine is Pulsatilla; then the Potency

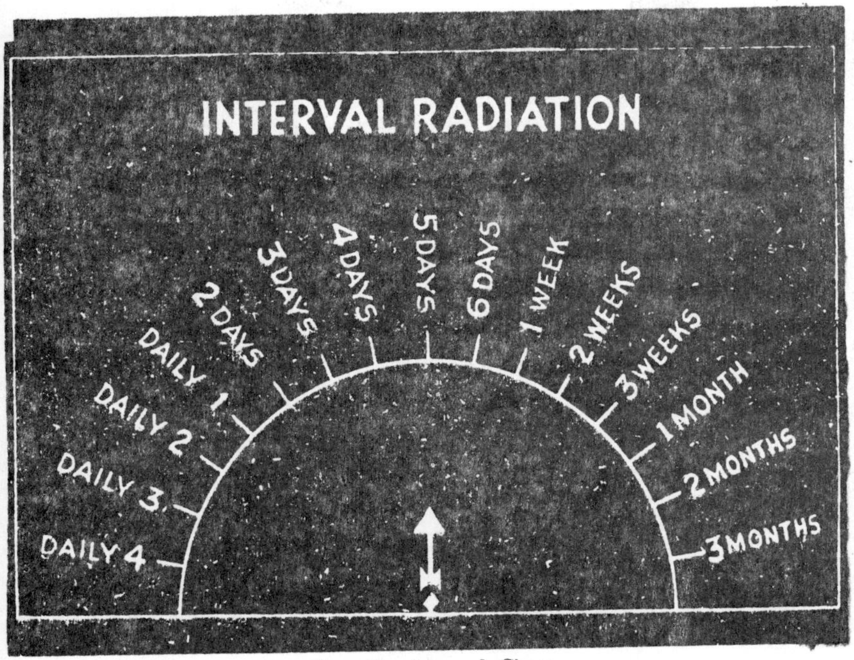

Fig. 40. Interval Chart

Radiation Chart is taken. On that the indicated potency is found out. Suppose, it is 1M; the next step is to dowse on the Interval Radiation Chart to find out the interval between two doses to be given to Mr. A. Suppose, it indicates 1 week; then this means that Pulsatilla 1M once a week may be given to Mr. A. After getting the above data it should finally be checked on the Percentage Chart for finding out whether it indicates 100%. When this indication is there, the medicine may be given with full confidence with the expectation of a good result.

The dowsing technique for the above tests should be followed scrupulously. As it has already been discussed previously, I am not repeating the details of the same.

CHAPTER IX

Magnet Study of Allied Sciences

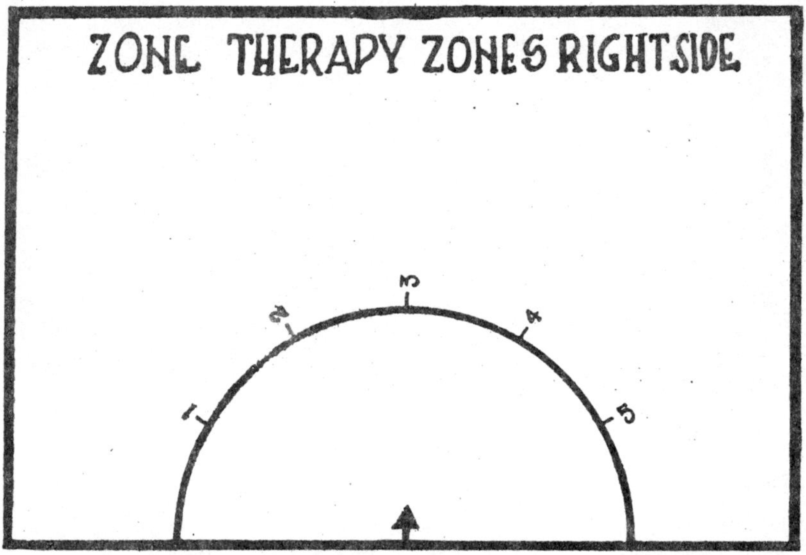

Fig. 41. Zone Therapy (Right Side) Chart

The science of Radi-esthesia or Magnet Dowsing may also be utilised for finding out many useful informations regarding medication, the way the medicines should be introduced in the system, the need for natural vitamins, etc. In this chapter I am including some more charts, which may be of use to practitioners and laymen alike to give relief to suffering humanity.

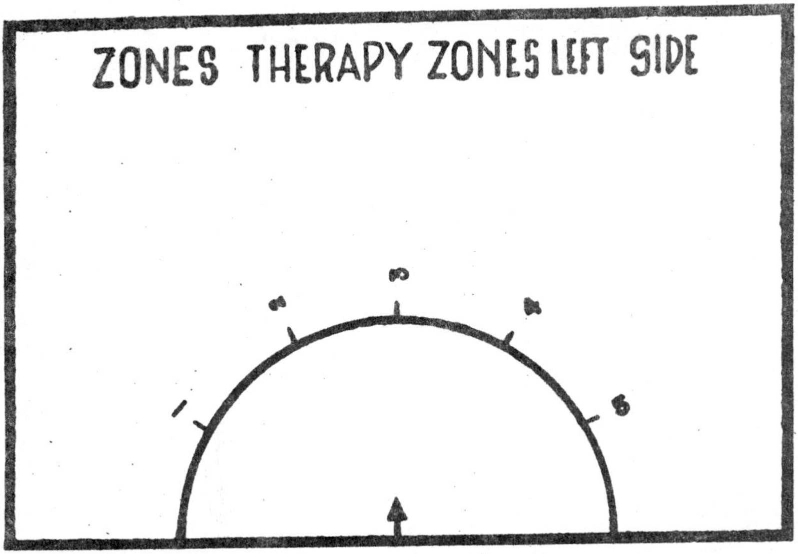

Fig. 42. Zone Therapy (Left Side) Chart

The first two charts are the Zone Therapy charts. One of the right side is Figure No. 41 and the other of the left side is Figure No. 42. The basic principles of Zone Therapy are as follows. The whole body is divided into ten sections—five on the right side and five on the left side. These zones are traced from each finger of the hand to the corresponding toe of the foot. Thus, the zone from the thumb to the big toe passes through the various organs, such as, stomach, bladder, etc. Dr. Fitzgerald discovered that deep massage and pressure on certain zones of the body restores organic functional activity in all regions situated in the zones treated. Treatment is mainly applied to palms and soles. The two charts given here will help find the zones which are to be given the pressure and massage treatment. Readers interested in

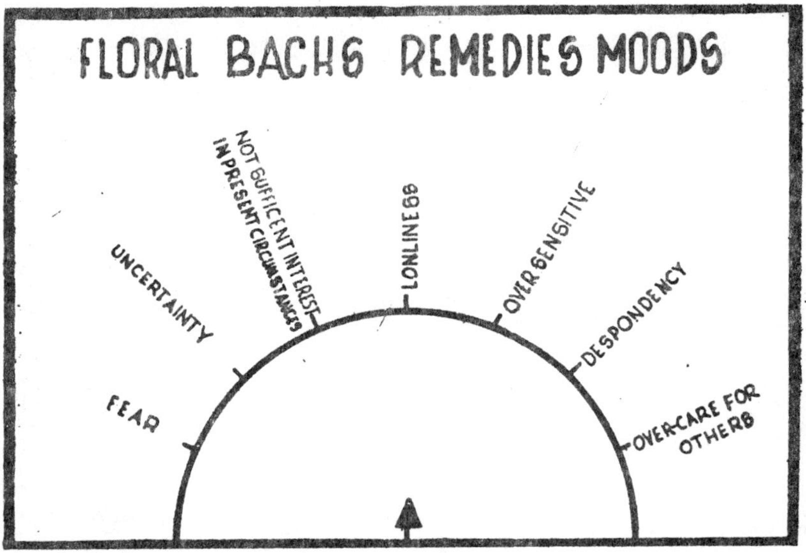

Fig. 43. Floral Bach's Remedies Moods Chart

knowing more about the subject of Zone Therapy may read "Zone Therapy" by Harry Bond Bressler, D.C., N.D. (Health Research, Mokclumne Hill, Calif., U.S.A.).

The next chart will be important to those who practice with the herbal or flower remedies discovered by Dr. Edward Bach of 'The Dr. Edward Bach Healing Centre, Mount Vernon, Sotwell, Wallingford, Berkenshire. There are 38 remedies, one for each of the main negative states of mind or moods common to mankind. The chart Floral Bach's Remedies Moods (Fig. 43) gives some of the important mental states. A patient's photograph may be dowsed on this chart and the mental state will be indicated by the pendule being attracted towards a particular entry. This will help in prescribing the Floral remedies of Dr. Bach.

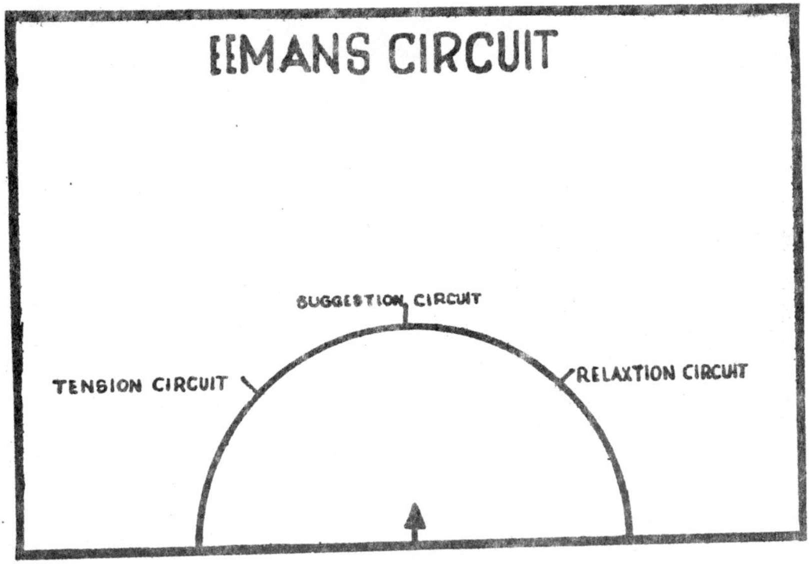

Fig: 44. Eemans Circuit Chart

The chart Eemans Circuit (Fig. 44) will be interesting to those who like to conduct research on the basis of the findings of late L. E. Eeman (Ref. "The Curative Properties of Human Radiations", 1947, Frederick Muller Ltd., London WC 1."). The basic principle is this 'when different parts of one human body or different or similar parts of different human bodies are connected by means of electrical conductors like insulated copper wires, these bodies behave as though—using an electro-magnetic analogy—they were bi-polar'. Any arrangement which connects polar opposites of one or of different bodies by means of electrical conductors is referred to as a "Relaxation Circuit" and any arrangement which connects polar similars as a "Tension Circuit". The chart

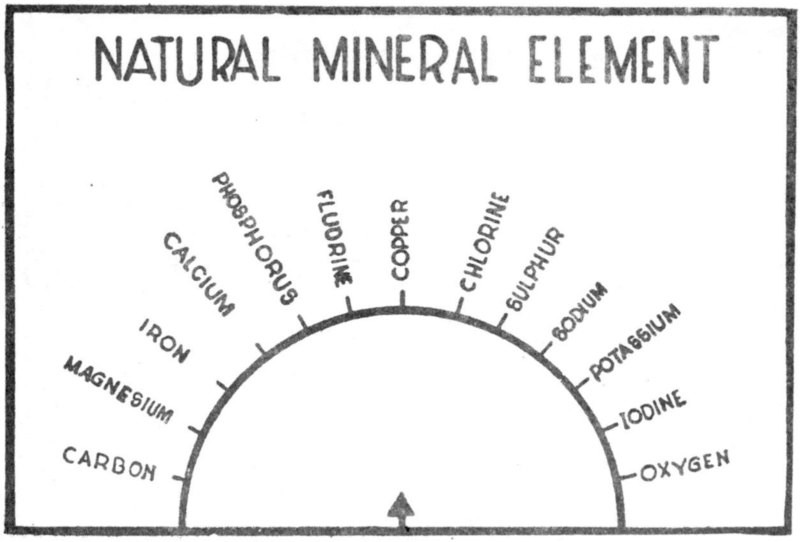

Fig. 45. Natural Mineral Element Chart

will indicate which circuit is needed for a particular patient when the photograph or the name card of the person is dowsed on the chart.

Along with any medicinal treatment Diet forms a very important factor in the restoration of health in a sick person. This diet is again composed of the natural minerals and the natural vitamins. The two charts given here are the Natural Mineral Element Chart (Fig. 45) and the Natural Vitamin Chart (Fig. 46). These two charts should always be consulted and the diet of the person should be fixed accordingly. The photograph should be placed at the base of the chart on the Magnet Box and dowsing should be done at the centre point. The pendule with a lower pointed end will point out the

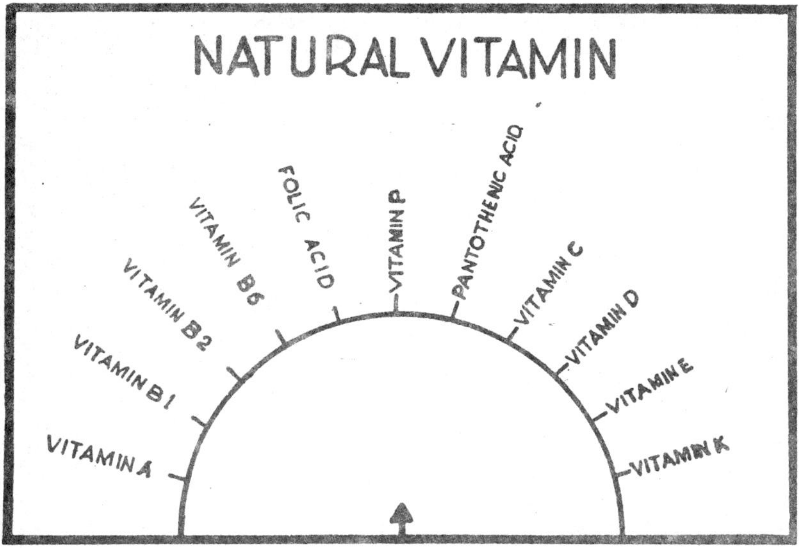

Fig. 46. Natural Vitamin Chart

minerals and vitamins needed by the person. Everyone needs these minerals and vitamins. All natural foods are rich in these elements. Vitamins should be taken as provided by nature. Synthetic, man-made substitutes should be avoided.

Only a few systems of healing have been referred to show that the art of Magnet Dowsing opens a vast field of research which can be carried out for other various natural systems of medicines.

The subsequent charts are meant for very advanced dowsers and it is earnestly hoped that there will be no misuse of the charts. These seven charts give the Chyle Medicines (Fig. 47), Blood Medicines (Fig. 48), Flesh Medicines (Fig. 49), Fat Medicines (Fig. 50), Bone Medicines (Fig. 51),

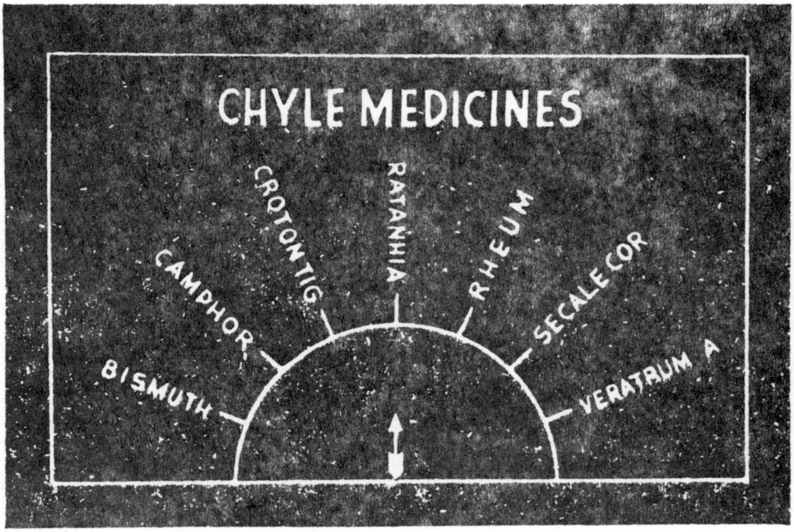

Fig. 47. Chyle Medicines Chart

Marrow Medicines (Fig. 52) and Sukra Medicines (Fig. 53). These charts are specially used for preparing Septenate mixtures either for specific diseases or for Personal Septenate Mixtures. This is a very laborious process and requires considerable time before a mixture can be prepared. The modus operandi is as follows :

For preparing a septenate mixture for a disease the name card of the disease is prepared or the pathological specimen is kept on the Magnet Box and dowsing is done on the first chart "Chyle Medicines". After careful dowsing the result should be noted down. Then the next chart "Blood Medicines" should be taken and dowsed in the above manner and the result noted down. In similar way the other medicines should be found out from the Flesh chart, Fat chart, Bone chart,

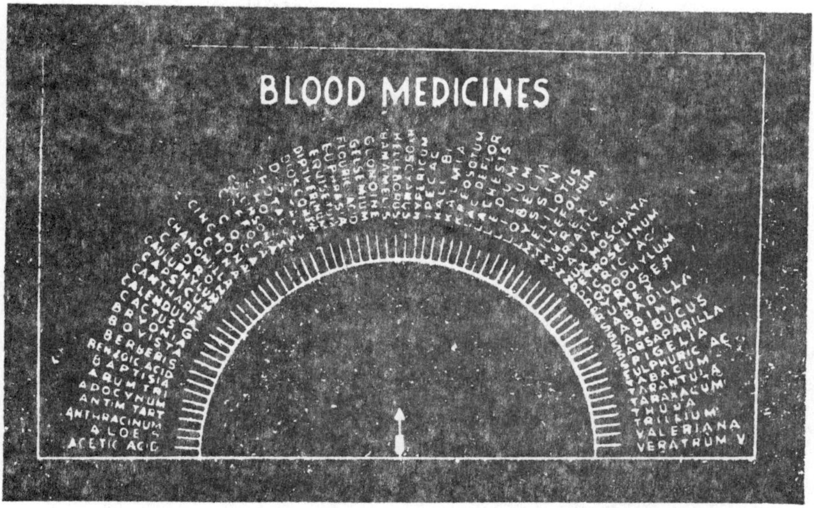

Fig. 48. Blood Medicines Chart

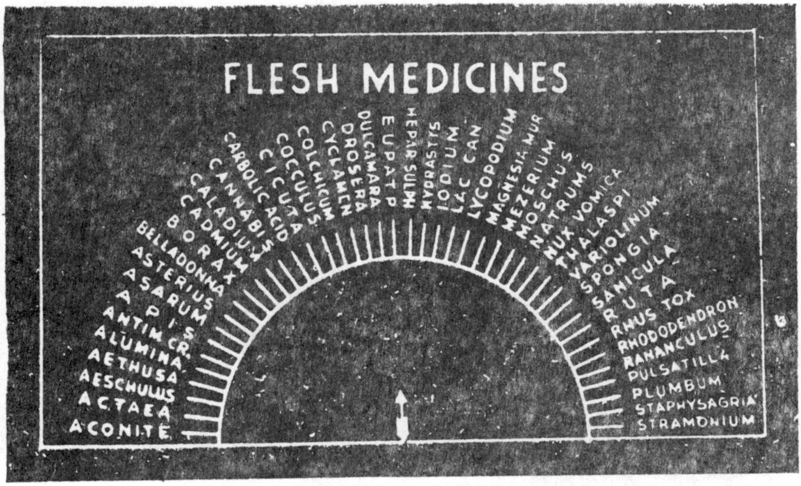

Fig. 49. Flesh Medicines Chart

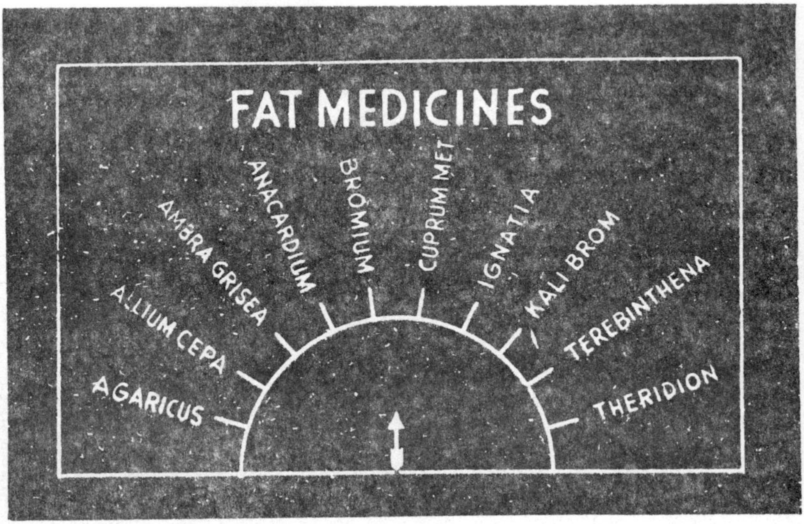

Fig. 50. Fat Medicines Chart

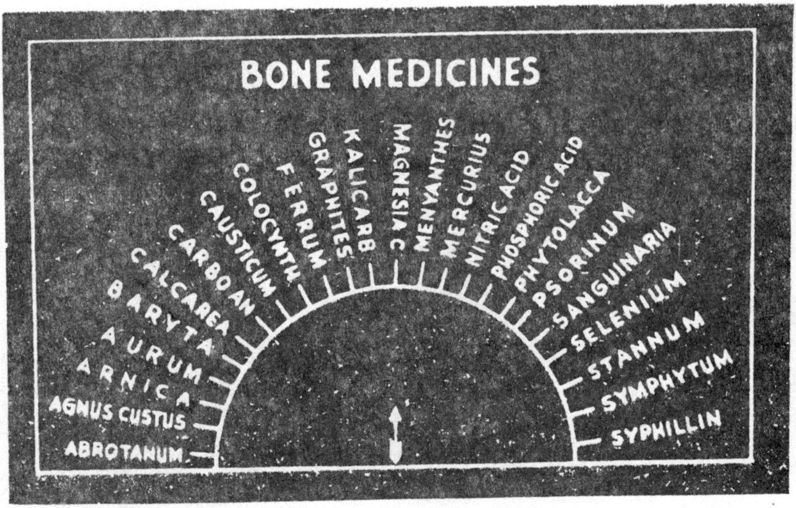

Fig. 51. Bone Medicines Chart

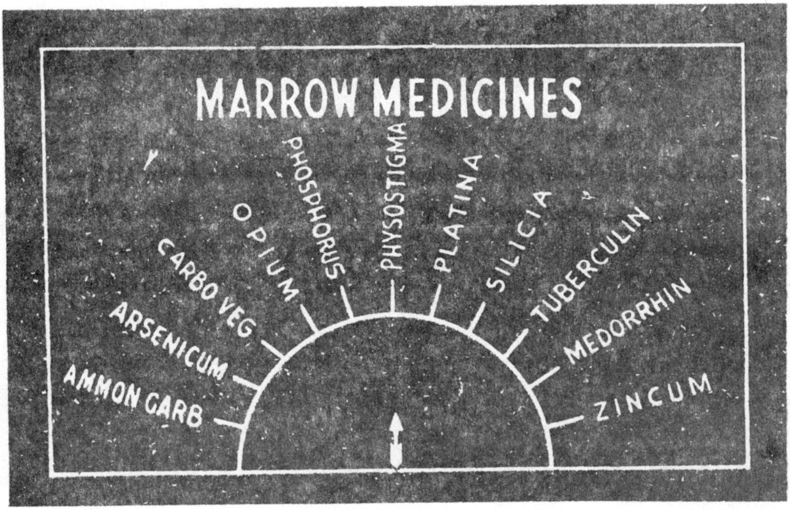

Fig. 52. Marrow Medicines Chart

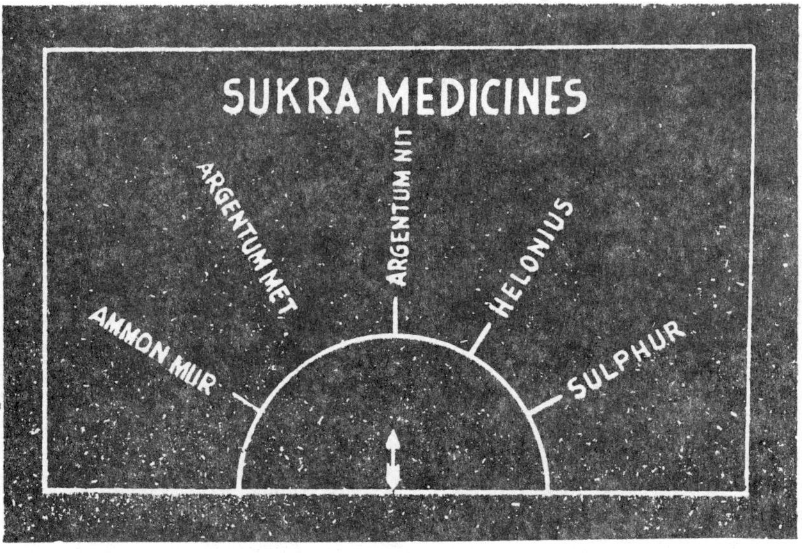

Fig. 53. Sukra Medicines Chart

Marrow chart and the Sukra chart. When all the seven medicines are found out it should finally be checked on the percentage chart to know whether they are 100% correct. In case they are not 100% correct, the dowsing should be done again with all the charts till it shows 100% correct. When the septenate mixture is prepared for a patient, the photograph or name card of the patient should be taken and the same procedure is to be followed. When the mixture is finally prepared and 100% accuracy is found, the Potency and Interval between two doses should be found out by dowsing on the respective charts. Once again, it may be stressed that it takes good deal of practice and patience to prepare a Septenate Mixture from the above charts by dowsing and great caution is to be observed.

CHAPTER X

Magnet Study of the Seven Plexuses

The seven plexuses are the seven centres of power in the human body for the operation of the seven cosmic rays, which represent the five elements, the five Vayus and the seven planets as listed previously. The plexus called the Manipura Chakra which is almost equivalent to the Solar Plexus of modern anatomy, is the centre from which the Red cosmic ray, the element of fire, the Samana Vayu and the planet Sun operate in the human system. In health and disease it is of utmost advantage to find out the offending plexus, and give a quietus to the famished power centre by a potentized medicine which has the power to strengthen the particular plexus or by radiating the respective cosmic colours through Teletherapy or Gem Therapy.

Similarly, the Cardiac Plexus or the Anahata Chakra is the power centre for the Violet cosmic colour, the element of air, the Udana Vayu and the planet Saturn. When this Vayu is involved it gives rise to several nervous disorders of a painful kind. It is profitable to know what particular plexus is deranged by disease, in order that the offending plexus may be given the required nourishment either through cosmic colours, wireless vibrations of gems, gem medicines or homoeopathic medicines having power over that particular plexus.

These and other questions are discussed in the present chapter, and they will be found to be of interest to all who are eager to study human maladies and their medicines. The importance of the seven plexuses or the Yogachakras as they are called in the Science of Yoga has not yet been recognised in any system of medicine. In Ayurveda elaborate description of the functions of the plexuses are not to be met with anywhere, but there is evidence to show that some of the great medical men were acquainted with the working of some of the Chakras (plexuses) in the human body. When for instance, they prescribe such tonics as Chyavanaprasa they seek to nourish the Muladhara Chakra which rules the heavy substances of the body such as flesh, fat, bones and the rest. The Muladhara Chakra being of the nature of the earth element integrates the material substances of the body. As far as I have been able to gather the Muladhara Chakra is a small little motor at the lowermost end of the spinal column which regulates in the body the colour currents of the Yellow cosmic ray. The colour Green being of the nature of the five Tattvas or Elements has also power over the Earth element principally, and over the other Tattvas secondarily.

The human body is compared to a microcosm and it contains all the forces the macrocosm has. The elements that are found in the macrocosm are present in the microcosm, but their parallels are not so markedly noticed in any science other than the Yoga system and the Tantras. Here an attempt will be made to trace the macrocosmic elements in the human body as revealed in the Yoga system. This information is of utmost practical value in life and particularly in the science of medicine.

The Brahman is said to be the life-giving principle in the macrocosmic world; it is co-extensive with the universe endowed with the three divine qualities of omniscience, omnipotence and omnipresence. In the Vedanta system of philosophy of India it is laid down in the very beginning: ALL THIS INDEED IS BRAHMAN. According to Vedanta, therefore, there is nothing tangible in the universe which is not part of Brahman, and which is not covered by Brahman. Brahman is everything, Brahman is everywhere and Brahman pervades everything. Now it is for the modern generation to see whether this is a scientific truth or mere philosophical nonsense.

Brahman is the macrocosmic soul. Its microcosmic parallel is the Jivatman or the microcosmic soul. The cosmic life principle is present in the Jivatman (individual soul). The Jivatman is the condensed form of the macrocosmic soul. Its place in the scheme of the Yogachakras in the human body is in the Sahasrara Chakra or the connecting gland of the four lobes of the cerebrum. Expert Ayurvedic physicians even at the time of death could excite this Chakra with their Hemagarbha Matra to bring back consciousness to a dying person. According to what I have learnt this Chakra is a microscopic motor which is run by the cosmic colour Green and seeks to coordinate the life current which is a totality of all the seven cosmic colour currents received from cosmos.

In the macrocosm again, there is a cosmic mind or cosmic intelligence which creates, maintains and destroys the universe along with its mobile and immobile contents. In the Yogachakras the place of the human mind or will is in the Ajna Chakra which lies over the topmost end of the spinal column. It is said in Yoga works that the fluid of immortality is constantly oozing out of the Sahasrara Chakra and is falling on the Ajna Chakra. With the help of this fluid of immortality life is sustained in the human body. This flow is stopped at the approach of death and life energy disappears from the body. This Chakra is called the Ajna Chakra or the Wheel of Command and it guides and controls all conscious activity of the human body. This Mind Chakra is of the nature of the Akasha Tattva or the primaeval Ether principle which animates everything from the minutest dot to the highest planet. According to what I have been able to gather this Ajna Chakra is the microscopic motor which is operated constantly by the cosmic colour Blue which rules over the Gland and the Fat system and on the life principle generally.

The process of evolution proceeds further with the production of the five primaeval elements in the human body as they are found in the macrocosm. The elements are evolved in a particular order, and this order is stated in the Upanishads thus: In the process of evolution, the element of Akasha or Ether comes first, it is followed by the element of Air or Vayu. Air in its turn is followed by the element of Fire (Agni). Fire gives rise to Water (Ap) and water in its turn produces Prithvi or the element of Earth. The passage giving this origin of elements constituting the universe is the briefest possible, and is quoted below:

Akasad Vayuh	(Vayu from Akasa)
Vayor Agnih	(Agni from Vayu)
Agner Apah	(Water from Agni)
Adbhyah Prithvi	(Earth from Water)

In the Yogachakras these elements have been assigned their distinct positions. Their place is deep inside the central core of the spinal cord and they are of the shape of small microscopic lotuses. The Chakras, according to Yoga works, manifest themselves one after another in evolution, and they disappear at the time of death in their reverse order in involution. Thus after death it is not possible to discover these Chakras or lotuses by any kind of post-mortem examination, even as the soul cannot be discovered in the human body by any kind of surgical operation.

The element of Akasha has its centre of power in the Ajna Chakra and it rules all empty spaces in the human body. Sound is the quality of Akasha (ether) and it is guided by this Chakra. So also the ears with which we hear sounds. In astrological works this function is given to Jupiter who is the ruler of the Gland or Fat system and is the lord of speech. There is still some controversy on the point as to whether ether should be connected with the Ajna Chakra or the Visuddhi Chakra or with both. In astrological works the gland and the fat systems are definitely under the sway of Jupiter, the Lord of the Ajna Chakra.

The Visuddhi Chakra comes next in order, and it is of the nature of a microscopic motor which regulates the action of the Indigo cosmic colour received from cosmos. This Chakra keeps up the purity of the organism intact by keeping it free from invading germs and bacilli and maintains the health of the thick lymphs in the body and the sperm system.

The element of Air has its centre of power in the Anahata Chakra below the Visuddhi Chakra. This Chakra is given the name of Anahata because it gives the heart sounds without being struck. This Chakra rules over the entire skin system and the sensation of touch. The Anahata Chakra is akin to the Cardiac plexus of modern anatomy. It is ruled over by the powerful planet, Saturn who is also the lord of all nerves in the body and contributes the Udana (Upward) Vayu which when deranged gives rise to fearful Vayu diseases like fits of unconsciousness, Insanity, Epilepsy, and Asthma. The

Anahata Chakra, as I have understood, is like a microscopic motor which receives from the cosmos the Violet cosmic colour and regulates its movement throughout the body. Much of this knowledge is derived from the magnet and authorities are welcome to differ from these views. This applies equally to the remarks on other Chakras also.

The element of Fire comes next in order and its centre of power is in the Manipura Chakra. Its rulership is over the general heating system of the human body, and its special function is to comprehend all kinds of forms through the sense-organ of eyes over which it therefore presides. Sight is said to be the quality of the element of fire in the human body. The Manipura Chakra is usually identified with the Solar plexus or the plexus of the coeliac-axis of modern anatomy. According to what I have been able to understand the Manipura Chakra is a small little motor which receives the Red Cosmic colour from the cosmos and regulates its flow and action throughout the body.

The element of Water comes next in order, and its seat of power is in the Svadhisthana Chakra. It rules over all the thin lymphs of the body such as chyle, blood, serums, exhudations and the rest. With this power it is possible to comprehend different tastes with the tongue over which the Svadhisthana rules. In the Samkhya system Taste is given as the quality of the element of water. The Svadhisthana Chakra is sometimes identified with the hypogastric plexus of modern anatomy. According to what I have been able to understand the Svadhisthana Chakra is a small little motor which receives the Orange cosmic colour from the cosmos and regulates its flow and action throughout the human body and guides the movement of the element of water.

The element of Earth comes next in order and it rules over all the solid substances of the human body. Its seat of power is in the Muladhara Chakra. Its quality, according to the Sankhya system is Smell, and its chief sense-organ is the nose, with which we are able to comprehend smell. According to what I have been able to understand the Muladhara Chakra is one of the seven microscopic motors which receive from the cosmos the Yellow cosmic colour and regulate its flow throughout the human body, and thereby protect the earth element of the body and guide its movement. This colour has much in common with the Green cosmic colour which partakes of the nature of the five elements combined and has connection with the nose and smell because it is possible to find out with the help of the prism that the nose tip receives the Green cosmic colour from the cosmos and thus contributes the power of smell to the nose. When Green is decomposed it is split into two colours: Yellow and Blue. The colour Yellow radiates from the earth, while Blue falls from above on the top of every tangible thing.

That these Chakras or plexuses were regarded as centres of power is well illustrated by the fact that the Yoga Shastra prescribes a presiding deity for every Chakra, barring the Sahasrara Chakra, or the seat of the Jivatman. The Shakti or the power in the Muladhara is called Dakini. Likewise the Shakti of Svadhisthan is Rakini, the Shakti of Manipura is Lakini, the Shakti of Anahata is Kakini, that of Visuddhi is Shakini, while the Shakti of the Ajna Chakra is Hakini. These six Chakras along with the Sahasrara Chakra, are like small motors guiding and regulating the equilibrium of the different elements and functions and controlling their even flow and working. Just as one motor may be employed for pumping water, a second for producing the kitchen fire, a third for turning the saw, a fourth for producing light and a fifth for running fans, even so these small Chakras like small and powerful motors are continuously controlling and guiding the different functions of the human body, regulating the even flow of the different elements and receiving the cosmic colours from the universe. What we call the life current is composed of these seven cosmic colour currents and the life energy continues and this we call Life.[1]

The importance of the Yogachakras and the cosmic colours is being made apparent at every step, and the fairly accurate knowledge of the Chakras will prove a blessing not only to all who seek knowledge of self, but also to numerous people afflicted with disease. By finding out the defective Chakra and then correcting it with the appropriate cosmic ray the key will be acquired to the cure of such terrible diseases like T.B., Cancer, Insanity and Epilepsy, which are the despair of the doctor.

But the chief problem is how to find out what particular Chakra is out of equilibrium in a disease and requires nourishment, and unless this is known treatment becomes difficult, nay impossible. Sometimes I wonder how without this valuable equipment people can be treated with blind prescriptions. In finding out the defective Chakra Radi-esthesia or Magnet Dowsing can be of immense help. But it is doubtful whether many would take to it and employ it for the good of themselves and their patients, because of the time spent or the labour involved.

---

1. For a brilliant description of the seven plexuses, the Five Vayus and their working in the human body see THE MYSTERIOUS KUNDALINI by Dr. Vasant G. Rele, seventh edition, Bombay, Messrs D. B. Taraporvala and Sons.

Anyway, there is a good method for finding out the defective Chakra in individuals suffering from disease. A brief outline of the method is given here which may be given a trial by sincere investigators. A pendulum suspended with a black string about 6″ inches long should be held in the right hand, and with the left hand the different portions of the spinal column and the head of the patient should be touched. While touching one or another of the seven Chakras, the movement or rather the gyration of the pendulum should be carefully watched. If the Chakra gives a right to the left (clockwise) gyration then the Chakra should be regarded as healthy, and therefore it should not be interfered with. If however the gyration of the pendule is from the left to the right (anti-clockwise), then it should be concluded that it is unhealthy and requires nourishment. This nourishment can be given in various ways: by medicine, by cosmic rays, by electricity, or by magnets themselves.

When the health of the Sahasrara Chakra is to be investigated, the left hand should be placed on the top of the head, and the motion of the pendule in the right hand should be observed. If it gives right to left (clockwise) gyration it is healthy. For the Ajna Chakra the hind part of the head or the place between the two eyebrow should be touched with the left hand. If the pendule in the right hand gives off a right to left gyration it is healthy. But if otherwise, it is unhealthy and requires nourishment.

For the Visuddhi Chakra the left hand should be placed on the neck. If the pendule in the right hand gives a right to left gyration, it is healthy. But if it gives the left to right gyration, then it is not healthy and requires nourishment. Similarly, for the Anahata Chakra the spine just behind the heart region should be touched with the left hand, and the movement of the pendule in the right should be watched. If it gyrates from the right to the left, it is healthy. But if otherwise, it is unhealthy and requires correction.

For the Manipura Chakra that part of the spine should be touched with the left hand which is behind the naval in a straight line. Then the movement of the pendulum in the right hand should be observed in the aforesaid manner. If the gyration is from the right to the left, the Chakra is healthy. But if otherwise, it is not healthy and requires nourishment. For the Svadhisthana Chakra the left hand should be placed between the Manipura Chakra and the end part of the spinal column. The pendule as usual should be in the right hand, and its movement should be interpreted in the aforesaid manner. For the last or the Muladhara Chakra, the end part of the spinal column should be touched with the left hand and the movement of the pendule in the right should be watched in the manner aforesaid. If the pendule gyrates in the right to left direction, then it is healthy, but if otherwise it is unhealthy and requires nourishment.

The same experiment can be conducted with the help of the Percentage Chart to find out the defective Chakra, particularly for those who find it difficult to understand and interpret the peculiar movement of the pendule, known as 'gyration'. When the Percentage Chart is used, the Chakras should be touched with the left hand, one after another in the manner aforesaid, and its Percentage tested with the right hand on the Magnet Box. If the percentage is shown to be 100% the Chakra is healthy, but if it is less it is unhealthy and requires nourishment. There are pictures of the Chakras in the different parts of the human frame in books and in

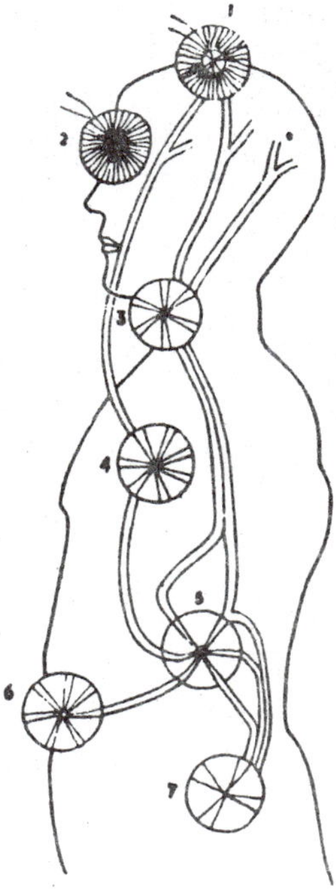

Fig. 54. Position of Seven Plexuses. (1) Sahasrara, (2) Ajna, (3) Vishuddhi, (4) Anahata, (5) Manipura, (6) Svadhisthana and (7) Muladhara.

separate sheets, and they are sold in the market (Fig. 54). If one can be secured for the purpose of test, the matter will be easy. The photograph of the person under test should be placed on the Percentage Chart, and with the left hand the different Chakras in the picture may be touched. And then by operating the pendule the percentage of the different Chakras should be determined. If the percentage is shown to

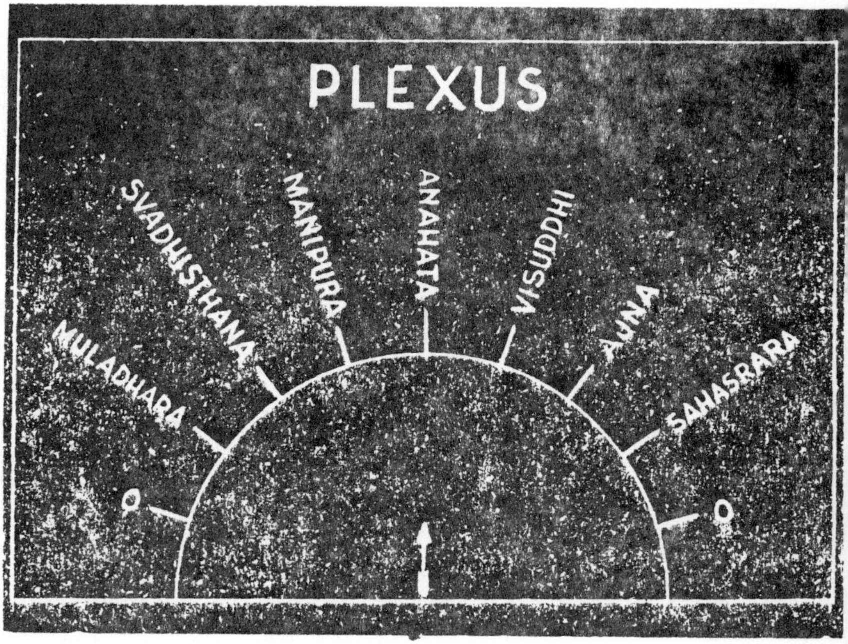

Fig. 55. Plexus Chart

be 100%, the Chakra is healthy, but if otherwise, it is unhealthy and requires nourishment.

To make things easy magnet dowsing on the Plexus chart may be tried. On the Plexus chart illustrated in Fig. 55 a photograph of the person under test should be placed on the centre of the base line, and with the right hand the pendule should be operated as before. By carefully watching the movement of the pendule the correct plexus to be nourished may be detected. By oscillation the pendule will go to one direction and point to the Chakra that is out of equilibrium. This experiment should be repeated several times and if every time the pointer goes to one particular entry on the chart, then the indication may be taken as fairly correct.

The seven Chakras mentioned in the Yoga and Tantra systems of India are somewhat vaguely described as plexuses in modern anatomy. Here the corresponding anatomical names are given with the proviso that these names are only approximately correct, but not exactly correct. From the well-known work of Dr. V. G. Rele: *The Mysterious Kundalini* published by Messrs D. B. Taraporewala & Sons, Bombay, these anatomical parallels of the Chakras are given here. The plexuses in modern anatomy are substances but are not regarded as centres of power as they are understood in the Yoga and Tantra systems.

The Muladhara Chakra is given the name of Pelvic Plexus in modern anatomy. The Svadhisthana is known as the Hypogastric Plexus, while the Manipura Chakra is called the Plexus of the coeliac axis or the Solar Plexus. The Anahata Chakra is called the Cardiac Plexus, the Visuddhi is called the Pharyngeal Plexus, the Ajna Chakra is given the name of the Cavernous Plexus (Nasociliary extension), while the Sahasrara Chakra is identified with the Cerebrum. As these anatomical positions are known, it would be possible to place the left hand on the proper spots for the different Chakras while testing them with the pendulum in the right hand or with the Percentage chart.

The seven Chakras have their connections with the five principal elements such as earth, water, fire, air and ether which are the five primordial constituents of the universe. Each cosmic element contributes one kind of Vayu or power both to the macrocosm and the microcosm. Earth contributes the Prana Vayu, Water contributes the Apana Vayu, Fire, the Samana Vayu, Air, the Udana Vayu and Ether, the Vyana Vayu. These Vayus are described in my Science of Tridosha published from New York by Messrs Gotham Book Mart. The Life Force is complete with these five Vayus and the elements that contribute them

The Muladhara Chakra controls the earth element in the body, the Svadhisthana Chakra the water element, the Manipura Chakra the fire element, the Anahata Chakra the air element, and the Visuddhi the ether element. The Ajna Chakra is considered to be the Mind of the individual, while the Sahasrara represents the microcosmic soul. People with a defective Muladhara suffer from constipation, diarrhoea, vomits and the rest. Those who have a defective Svadhisthana suffer from insomnia, mild form of insanity, stones in the kidneys and gall bladder, and various other diseases. A defective Manipura gives rise to mild asthma, rickets, arthritis, piles, and various other diseases. One who has a defective Anahata gets all kinds of nervous disorders, including fits of unconsciousness, neuritis, neuralgia, hysteria, epilepsy, severe asthma, deep-seated insanities, and similar diseases. An under-nourished Vishuddhi Chakra gives various ailments such as pus, boils, swellings, tumours, diabetes, granular lids, stye, and diseases of that kind. The Ajna Chakra, when defective, gives rise to various complaints such as swellings due to water accumulation, diseases arising out of germs and bacilli, diseases of the bronchii, lungs, heart diseases and palpitations. When the Sahasrara becomes defective life principle is at the lowest ebb. In all chronic and acute diseases when the patient is on the verge of death the Sahasrara Chakra is bereft of nourishment. In dangerous and killing diseases like typhoid, T.B., cancer, small-pox, measles, blood pressure, and the rest, the Sahasrara loses power, and is nourished quickly by the radiation of Green ray.

The seven plexuses again have connection with the seven Dhatus (systems) of the human body, such as, Chyle, Blood, Flesh, Bone, Marrow and Sperm. Chyle and Blood being watery in nature come under the Svadhisthana Chakra, Flesh is under the Muladhara Chakra, Fat including glands is under the jurisdiction of the Ajna Ghakra, Bone is under the jurisdiction of the Manipura Chakra. The Marrow system is under the jurisdiction of the Anahata Chakra and partly under the Muladhara Chakra, while the Sperm system is under sway of the Vishuddhi Chakra. The Sahasrara is of the nature of the five primordial elements collectively, and controls everything in the human body. The Yoga system advocates the piercing of the six Chakras by certain Yoga processes, and it is recommended that no attempt should be made to pierce the Sahasrara Chakra. As far as possible this Chakra should not be interfered with by mental exercises as prescribed in the Yoga system, but medicine which can influence the Sahasrara can be used both as a tonic and as a cure. But this should be done with caution and all possible care.

Finally, it is necessary to repeat and indicate here the cosmic colours which have power to nourish the different Chakras when they are undernourished, and by this weakness give rise to various diseases. For the sake of convenience the information is given here in a tabular form thus:

Chakra	*Cosmic Colour*
Sahasrara	Green
Ajna	Blue
Vishuddhi	Indigo
Anahata	Violet

Chakra	*Cosmic Colour*
Manipura	Red
Svadhisthana	Orange
Muladhara	Yellow

The subject of Chakras and their cosmic colours is a highly controversial one. Here the views have been expressed after rigorous tests with the magnet. At places there may be reasonable doubt, but just for the present the colours noted here may be taken as good enough for practical purposes.

CHAPTER XI

Advanced Dowsing

From the foregoing it will be apparent that dowsing in the simplest form requires practically nothing—a small little pendulum made of anything metal, glass bead, agate, crystal or even wood. With this simple contrivance, called the pendule, all kinds of tests can be made. For instance whether a particular food is agreeable or beneficial to a person or not can be found out easily by this process. A dish containing say, a piece of wheat bread, is placed on the table before the operator. Between the dish and the operator the name card of the person or his photograph or signature may be kept, and over this card the pendule may be operated. The pendule should be given first of all, a gentle rotatory movement from the left to the right, and then allowed to move by its own force. It will either go towards the dish of bread or will move parallel to the dish. If the pendule is attracted towards the bread, then bread is beneficial to the person for whom the test is undertaken. But if it is repelled by running parallel to the dish, then it must be concluded that the bread is injurious to the person as a diet. After repeated tests the decision should be made, and a little practice in this line gives to the operator the required mastery.

All doctors can make use of this simple device to find out whether a medicine fixed for the patient will be beneficial or not. If the pendule moves towards the phial or the tube containing the medicine, then it may be given. But if the pendule runs parallel to the medicine it would be advisable not to use it for the patient under test. This simple test is recommended in all earnestness to all Homoeopaths, and they will be surprised to find how many incorrect prescriptions are made by them. The test can be made still more simple. The phial selected should be placed on the table, and the name of the patient should be mentally repeated and then a gentle rotatory movement from the left to the right should be given to the pendule as before. The pendule will run towards the phial if it is the correct medicine, otherwise it will run parallel if it is wrong. The correct medicine should be prescribed and the wrong medicine should be rejected. A little practice will teach the secrets of selection and will make the physician an expert prescriber. The above method is known as Psychic Dowsing and when one has practised ordinary dowsing for a long time as advocated in this book, this method may be tried when there are no charts or photographs available.

The same experiment can be done in a different way. The medicine phial for a particular patient may be placed on the table, and while mentally repeating the name of the patient the pendule should be operated over the cork of the phial. If the pendule develops an up and down movement then the medicine is right and can be prescribed. But if the pendule runs parallel to the cork, the medicine is wrong. The right medicine should be prescribed, while the wrong medicine should be rejected.

Dowsing at a higher stage requires higher magnetisation. Here the chart has to be magnetised, and the operator's body and mind have to be magnetised. If the chart is magnetised it is not necessary to hold a magnet as a pointer. When dowsing on the Electro-Magnet Box it is not necessary

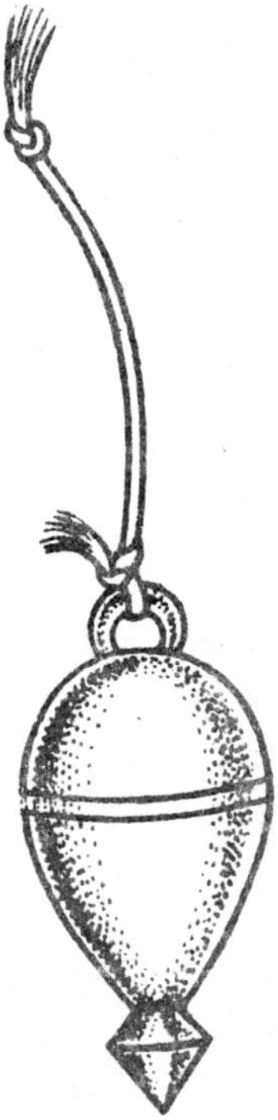

Fig. 56. Silver Pendule

to place a magnet on the chart as the chart is already placed on highly dynamic magnetic field.

A pointer of silver or crystal or glass with a pointed lower end (Fig. 56) will be necessary. The foregoing charts for the purpose of dowsing can be used only under this arrangement. Constant practice is the only other thing required to gain proficiency in the art of magnet dowsing. In this form of dowsing only two magnets will be necessary ; one of the two should be used as a magnet pointer. The other magnet should be held in the left hand to magnetise the body and mind of the operator. This magnetisation of the body and mind is important because dowsing is affected if there are strong prejudices in the mind of the operator either in favour or against a particular answer. If a magnet is held in the left hand whatever may be in the mind, subconscious or conscious, it will not affect the result of tests. The magnet that magnetises the body and mind should be a power 'C' eclipse magnet or any other of equivalent strength, to be efficient. There is no general rule with regard to the lifting power of the magnet, but practical experience in the line shows that the magnetisation is efficient, if the magnet is powerful.

Still higher magnetisation is necessary in order to obtain more accurate results. Many experiments were done with magnets arranged in different ways in a box, which were known as Magnet Boxes. The first Magnet Box that was prepared contained 8 to 16 magnets arranged in a particular manner. All the magnets were first examined for polarity and their North poles were marked. The magnets were then arranged at the bottom of the box in two lines if 8 magnets were used or 4 lines if 16 magnets were used. The magnets were attached to iron plates to keep them secured in place. A flap was then covered over the magnets and dowsing was done on this flap. The arrangement of the magnets and their poles in the box of 16 magnets was as below:

SN	SN	SN	SN
SN	SN	SN	SN
SN	SN	SN	SN
SN	SN	SN	SN

57. Magnet Box with 16 Magnets

The strength of the magnet box then depended on the lifting power of the magnets. If the magnets could lift a weight of 2 lbs. then the lifting power of the box came to 32 lbs. of weight. The Magnet Box is shown in Fig. 57. When the above Magnet Box was not found very useful another one was prepared and here the arrangement was entirely different. Twentyfour magnets were used and as in the above case the North pole was marked in the beginning. The box had three sections. In the left hand section 12 magnets were arranged with the North pole facing the top of the box and in the right hand section the other 12 magnets were arranged with their South pole facing the top. All the sections were covered with

Fig. 58. Improved Permanent Magnet Box

a flap and dowsing was done on the middle flap, this being the neutral area. The photograph was kept either in the left or in the right hand section and dowsing was done on a chart kept in the middle. Similarly, instead of the chart, if affinity was to be found from amongst different gems or medicines, they were arranged on the centre portion and dowsing was done there. To find out the affinity between a person and a particular medicine or gem, the photograph and medicine or gem were kept on either side of the magnet box and dowsing was done in the middle. If the pendule moved up and down, it indicated that there was affinity, whereas, if it moved parallel or horizontal, it showed that there was no harmony and the medicine or gem was rejected. The instrument is shown in Fig. 58.

The next step in the experiment for the dowsing box was the introduction of electromagnets instead of permanent magnets. The new instrument consists of a wooden box inside which there are three electromagnets arranged in a line equidistant from each other, nine bulbs and their individual switches, and a 2-volt transformer. There is a partition of wire mesh which covers the box except the switches and bulbs which remain uncovered. There are three sections on the wire mesh of the partition and under each section one electromagnet is placed. The electromagnets are connected in a parallel connection and the nine bulbs are connected with the transformer which again is connected with the main line. The wire which is to be connected with the main line comes out from behind the box. Outside, in front of the instrument is the master switch which operates the machine. The instrument works on 230 volts, A.C. current.

The nine bulbs represent the nine cosmic colours or the nine planets in the following order to be read from left as :

Ultraviolet-Violet-Indigo-Blue-Green-Yellow-
Orange-Red-Infra-Red

The planetary arrangement from left being—
Rahu-Saturn-Venus-Jupiter-Mercury-Mars-Moon-Sun-Ketu.
(Rahu and Ketu are the nodes of the moon).

To find out the colour hunger of a person, the photograph, the signature or any specimen of the person such as, blood, sputum, saliva, etc. is to be kept on either section of the wire mesh and the pendule is to be operated in the middle section. Before the operation the current should be switched on which will light all the bulbs. The question regarding the colour hunger may be repeated mentally and when the pendule starts moving, it will take a course towards a particular bulb which then should be identified according to the above

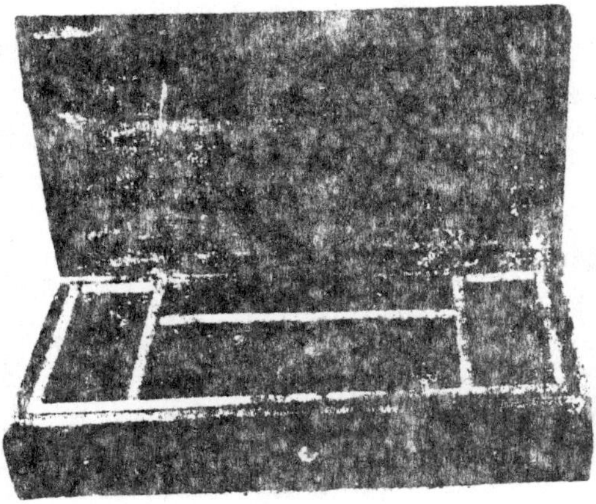

Fig. 59. Electro Magnet Box

arrangement and the colour will be known. To verify the result all the bulbs except the one indicated should be switched off and dowsing should be done to find out whether the colour indicated by the bulb was correct or not. If the pendule moves up and down, the colour is correct and if it moves parallel with the electromagnets, then the colour is not the correct one. Similar tests should be done by lighting the other bulbs till the correct colour is found out. In the case of dowsing on the electromagnet box, the pendule may or may not be given a slight clockwise motion. The Electro-magnet Box is shown in Fig. 59.

When dowsing is to be done for finding out other information about the person, all the bulbs should be switched off and dowsing should be done in the ordinary way keeping the photograph of the person either in the left or right hand section of the wire mesh. For example, if a harmony is to be found between the person being tested and a medicine, then the photograph and medicine should be kept in each of the sections and dowsing should be done in the middle. If the pendule moves up and down, the harmony is present but, if it moves parallel with the electromagnets, then there is disharmony. Similarly, when dowsing is done on different charts, they should be kept in the middle section and the photograph should be kept on either of the sections of the wire mesh. The lights should be switched off during the dowsing on the charts.

Ordinarily, dowsing on the electromagnet box is done with a magnet pendule but any other pendule will also serve the purpose. The advantage with the magnet pendule is that when it is held over the electromagnet, a flux is felt in the fingers, which shows that the machine is on. Personally I feel that this flux helps to keep the attention of the operator.

Still higher form of dowsing can be done by employing two magnet boxes and one 100 CM rule for greater accuracy and fineness of test. The two ends of the 100 CM Rule should bear two plastic containers glued to the wood for keeping various objects for the purpose of comparison and dowsing. The two magnet boxes should be placed on the table at such a distance that the two ends of the Rule may fall on the middle of the two magnetised flaps of the magnet boxes. When this is done, the two ends of the Rule and the Rule itself are fully magnetised. A small mark should be inscribed just in the middle point of the Rule where the magnet pointer should be operated. (Fig. 60).

For Rule Test of all kinds the operator should magnetise himself by holding a magnet in the left hand. Then various tests may be made by the Rule Method on the instrument for

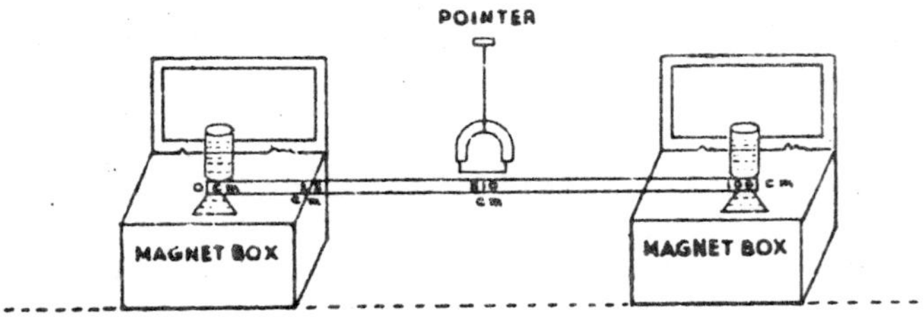

Fig. 60. 100 Cm. Rule on two Magnet Boxes.

advanced dowsing. For instance, in one of the containers a small gem, say Ruby, may be placed, and in another the photograph of a person or his name card may be deposited in order to see whether there is harmony between the gem and the name. Then with a small and light pendule or with a magnet in the right hand harmony or otherwise between the two should be tested. The pendule should be held over the central point of the 100 CM Rule at approximately 50 CM point. Then a gentle rotatory movement should be given to the right hand pendule from the left to the right, and it should then be allowed to move by its own force. If the movement is parallel to the Rule there is no harmony between the gem and the name. But if the pendule is thrown out of the Rule and it moves up and down then it is to be concluded that there is harmony between the gem and the person, and that if vibrations of Ruby are given to the photograph of the person, he will be benefited.

When very high and subtle forms of tests have to be conducted, this Rule method will be found to be of great value, and for this we have to thank the inventor of the Rule method, Mr. Wethered[1] of England. For instance, when from cards bearing names of diseases their cosmic colours have to be divined, this Rule Test is absolutely necessary. Similarly, when the cosmic colour inherent in Homoeopathic medicines has to be divined, this Rule Test is the only test that is applicable.

On one of the two containers one Homoeopathic medicine may be placed. This should be preferably in the left container. In the right container one after another the seven or even nine gems may be kept for test. In the middle point at about 50 CM the pendule should be operated and its movement should be observed with care. If it runs parallel to the Rule in several tests, then the gem should be rejected along with the colour represented by the gem. Any one of the seven or nine gems giving the right reaction to the pendule by its up and down movement will be the right gem, and its colour will be the right colour. It is in this manner that the cosmic colour of all new medicines can be fixed. Such tests are extremely delicate and should be done by very experienced dowsers, or if possible, by a society of excellent dowsers.

Omniscience that resides in the magnetic field can be exploited in various ways, and several methods have been clearly indicated here. There may be many more methods about which we know so little. So far horse-shoe magnets and electro-magnets have been used for the production of stronger magnetic fields. Still later, I believe, a time will come

1. In the Chapter on the Rule Method (pp. 39-47) Mr. Wethered the author of RADIESTHETIC APPROACH TO HEALTH AND HOMOEOPATHY published from England, has made a definite and far-reaching contribution to the science and practice of Radiesthesia. This Rule Method has been adopted here with the necessary modifications. In this book it is felt that the two unprotected ends of the 100 cm. rule should be magnetised with two magnet boxes at the two ends.

when the whole process will be mechanised, and it will be possible to exploit the omniscience that resides in the magnetic field by pushing only a button. We must wait till then for a fully mechanised dowsing system. Present dowsing will then be called scientific dowsing.

Finally, before closing, a few words may be said about the entries on the charts, and what they represent and how they respond to dowsing by the magnet or the pendule. In the higher cosmic field, it may be noted, there is no difference between Name and Form. If an actual Ruby is not available for the purpose of test the word "Ruby" may be written on a card, and this will represent the gem Ruby. Although the card bearing the Ruby cannot be cashed, its quality otherwise remains intact. The card will give the same vibrations as the gem Ruby, because the set of vibrations given out by the gem Ruby are all present in the word Ruby. Thus there is no difference here between the Name and Form of Ruby. Of the letters R-U-B-Y each is composed of the seven cosmic colours in a given quality and quantity. The four letters together form into a cluster with a definite set of vibrations made up of the cosmic colours, their arrangement, quality and quantity.

This set of vibrations is different from the vibrations produced by another cluster of letters, such as S-A-P-P-H-I-R-E. These vibrations and their difference are detected and recognised by the magnet and its uncanny magnetism which detects super-sensory cosmic vibrations of harmony and disharmony. If a prism is applied while a letter is being written on a plain paper, it can be seen how the cosmic colours present themselves round about the letters and how they arrange themselves in a particular way to give shape to the letter with the quality and quantity of cosmic colours. When the letters are erased, the cosmic colours disappear in the same way, and this becomes evident when it is observed with a prism.

Each entry on a chart is a magnet cluster and a living thing with likes and dislikes. The pendule which works on the principle of magnetism, understands this and moves towards that entry with which the person or thing under test has harmony, rejecting all others with which it is not in harmony. Higher and higher magnetisation only secures greater and greater accuracy.

In conclusion it may be pointed out that Magnet Dowsing will prove useful to all those who have a desire to probe the secrets of the universe, to have a full view of life, and to obtain a glimpse of the future. The magnet is a repository of divine power, and therefore, it has the three divine qualities of omniscience, omnipotence and omnipresence. There is no reason why all the findings of the magnet will be incorrect ; on the contrary, a substantial amount of the dowsing results have been found to be correct and are bound to be correct. Those who will work with the magnet for their knowledge today will be hailed as the wise men of tomorrow. The scientists who will be experts in magnet dowsing today will be welcomed as the wisest scientists of tomorrow. Small Homoeopaths today with Magnet Dowsing will become great Homoeopaths of tomorrow. There is a great dearth of books on divination, and the present author hopes that by this publication a long felt want will be removed. A glimpse of the future, however, hazy and indefinite, will be of immense value to every honest inquirer in whose hands this book is placed with very tender feelings.

END

BIBLIOGRAPHY

1. TELETHERAPY published by Dr. A. K. Bhattacharya 1967.
2. MAGNETISM by David Shoenberg, published by SIGMA, Introduction to Science No. 12.
3. THE MATERIA MEDICA OF THE NOSODES by Dr. H. C. Allen published by Boericke & Tafel, U.S.A 1910.
4. RAYS AND RADIATION PHENOMENA by Dr. Oscar Brunler published by DeVorss & Co., U.S.A. 1950.
5. THE SEVEN KEYS TO COLOUR HEALING by Dr. Roland Hunt published by C. W. Daniel Co. Ltd., Essex, England.
6. THE CHAIN OF LIFE by Dr. Guyon Richards, 2nd. edition published by Health Science Press, England.
7. THE SCIENCE OF TRIDOSHA by Dr. B. Bhattacharyya, 2nd. edition published by Messrs Gotham Book Mart, New York.
8. RADIESTHETIC APPROACH TO HEALTH AND HOMOEOPATHY by W. D. Wethered.
9. COURSE ON BIOMAGNETIC SCIENCES by Dr. Albert Roy Davis published by Roy Davis Research Laboratory, Green Cove Springs, Florida, U.S.A.
10. HOW TO USE A PENDULUM by Anonymous,

published by Health Research, Mokelumne Hill, California.

11. VITIC published by Borderland Sciences Research Associates foundation Inc., Vista, California.

12. MYSTERIOUS KUNDALINI by Dr. Vasant G. Rele published by D. B. Taraporvala & Sons, Bombay.

13. ZONE THERAPY by Harry Bond Bressler, D.C., N.D., published by Health Research, Mokelumne Hill, Calif.

14. THE MEDICAL DISCOVERIES OF EDWARD BACH, PHYSICIAN by Nora Weeks published by The C. W. Daniel Company Ltd. Ashingdon Rochford Essex England.

15. THE CURATIVE PROPERTIES OF HUMAN RADIATIONS by L. E. Eeman published by Fredri-ke Muller Ltd., London WC I—1947.